BIOLOGICAL SCIENCES

FROM TOXICOLOGY TO BIODIVERSITY

The Editors

Dr. Prashanth Konkal completed his M.Sc. in Zoology and Ph.D. from Karnatak University, Dharwad. He cleared the KSET exam (2014) and received the (University Grant Commission - Research Fellowship in Sciences for Meritorious student) Senior Research Fellowship (SRF). He has published research articles in national and international journals and presented at various scientific conferences. Currently, he is the Assistant Professor & Head of Zoology at Shri Gavisiddheshwar College, Koppal. Additionally, he holds key academic positions, serving as Member and Chairman of the Board of Examiners (BoE) of VSK University, Bellary and Koppal University.

Dr. Channabasava, born in Giniwar, Karnataka, is a dedicated botanist and educator. He earned his B.Sc. and M.Sc. from Karnatak University, Dharwad, and a Ph.D. under Prof. H.C. Lakshman. He cleared K-SET and has held teaching roles at KRCE, JSS College, and Prarthana PU College. Currently, he is the Principal at SGVVT's Shri Gavisiddeshwar College, Koppal. He has published 25+ research papers, 18 book chapters, and four books. An active academic leader, he contributes to botany, toxicology, biodiversity, and environmental conservation.

BIOLOGICAL SCIENCES
FROM TOXICOLOGY TO BIODIVERSITY

Prashanth Konkal

Channabasava

2026

Daya Publishing House®
A Division of
Astral International Pvt. Ltd.
New Delhi – 110 002

ISBN: 978-93-5919-294-9

Published by : **Daya Publishing House®**
A Division of
Astral International Pvt. Ltd.
– ISO 9001:2015 Certified Company –
4736/23, Ansari Road, Darya Ganj
New Delhi-110 002
Ph. 011-43549197, 8130496929
E-mail: info@astralint.com
Website: www.astralint.com

Printed at : **Replika Press Pvt. Ltd.**

FOREWORD

A dynamic interaction of biological, chemical, and environmental elements shapes the complex web of life on Earth. "Biological Sciences: From Toxicology to Biodiversity Conservation" offers a broad compilation of research that crosses several disciplines, providing important insights into ecological and biomedical advancements in a time when scientific exploration is essential for sustainable living. Studies that look at the impact of pollutants on aquatic life, the medicinal potential of substances derived from plants, and the role of microorganisms in agriculture are collected in this volume. Assessments of biodiversity, avian population studies, and conservation initiatives for ecologically significant fauna are also highlighted. The book also examines green technologies, such as the synthesis of bio-inspired nanoparticles and their uses in medicine. This book is a resource for scientists, conservationists, and policymakers who are working to strike a balance between environmental preservation and human progress by combining research from toxicology, ecology, and biotechnology. It is my hope that this work will stimulate more research and significant contributions to the biological sciences.

Prof. Taranath T.C
Vice-Chancellor
Hasan University, Hasan

PREFACE

The biological sciences have seen remarkable advancements in recent years, spanning disciplines from toxicology and environmental stress physiology to biodiversity conservation. This book, Biological Sciences from Toxicology to Biodiversity, aims to bridge these fields, providing researchers, academicians, and students with a comprehensive resource on emerging trends and challenges. The chapters in this volume delve into various aspects of biological research, including the impact of environmental pollutants on organisms, adaptive responses to stress, conservation strategies for endangered species, and novel methodologies in ecotoxicology. By bringing together contributions from experts in these domains, we seek to provide a holistic understanding of the interconnectedness of life sciences and environmental sustainability. This book is intended for a broad audience, including biologists, ecologists, environmental scientists, and policymakers who are keen on understanding the latest developments in biological sciences. We hope that it serves as a valuable reference for both theoretical insights and practical applications in research and conservation efforts. We extend our deepest gratitude to all contributing authors, peer reviewers, and institutions that supported this endeavor. Their expertise and dedication have been instrumental in shaping this work. We also acknowledge the unwavering support of our colleagues and families, whose encouragement has been invaluable throughout this process. We sincerely hope that this book inspires further research and contributes to the advancement of biological sciences for a sustainable future.

Prashanth Konkal

Channabasava

CONTENTS

CHAPTER 1

EFFECT OF CYPHENOTHRIN ON THE HISTOARCHITECTURE OF GILLS, LIVER AND KIDNEY OF A FRESHWATER FISH, *CIRRHINUS MRIGALA*

Srinivas B Neglur[1], Sapna S Anigol[1], Muniswamy David[2*]

PG Department of Studies in Zoology
Environmental Biology and Molecular Toxicology Laboratory
Karnatak University Dharwad-580003

ABSTRACT:

The histopathological effects of cyphenothrin on the gill, liver, and kidney tissues of the freshwater fish, *Cirrhinus mrigala* were determined by light microscopy. The experimental fish were divided into groups of 10 (n=10) and exposed to different concentrations of the test pesticide for 24, 48, 72, and 96 hours. The 96-hour LC_{50} was found to be 30µg/l. The fish were exposed to sub lethal concentration of 6µg/L for 10, 20, 30 and 40 days to pyrethroid, typically known as Cyphenothrin (5% EC). The most common gill changes at all durations of Cyphenothrin were desquamation and necrosis. Besides, aneurism in secondary lamellae, lifting of the lamellar epithelium, oedema, epithelial hyperplasia, and fusion of the secondary lamellae were reported. The changes observed in the liver tissues were cloudy swelling, focal necrosis and hypertrophy of hepatocytes, degeneration of hepatocytes and cytoplasmic, extensive vacuolation of hepatocytes and pyknotic nuclei. Lesions in the kidney tissues of fish exposed to Cyphenothrin were characterized by degeneration in the epithelial cells of renal tubule, pycnotic nuclei in the hematopoietic tissue, dilation of glomerular capillaries, degeneration of glomerulus, intra cytoplasmatic vacuoles in epithelial cells of renal tubules with hypertrophied cells and narrowing of the tubular lumen.

Keywords: Cyphenothrin; Histopathology; Cir*rhinus mrigala*; Gill; Liver; Kidney.

Introduction

Water contamination from industrial, agricultural run-off, and municipal sources has become a major concern for human safety due to population growth, urbanization, and industrialization Yousfzai *et al.*, 2011. Pesticide use has skyrocketed over the past few decades all throughout the world. The ecosystems have become poisonous due to the overuse of pesticides. Pesticides are detected in groundwater, crop samples, and agricultural streams in varying concentrations, according to several monitoring programs and scientific research Mahboob *et al.*, 2015; Sharma *et al.*, 2015; Yadav *et al.*, 2015). The environment and surface waterways were contaminated with several insecticides, according to studies by Arjmandi *et al.* 2010 and Rahiminezhhad *et al.* 2009. The impact of pesticides on the health of aquatic creatures was documented by Banaee and Ahmadi 2011.

Pyrethroids are inferiors of normal Pyrethrins derived from the flowers of the pyrethrum plant (*Chrysan- themum cinerariaefolium* and *C. coccineum*) Gupta 2018. They include esters of chrysanthemum acid (ethyl,2- dimethyl- 3-(1-isobutenyl) cyclopropane-1-carboxylate) and halogenated inferiors of their acids and alcohols Costa 2015. Despite the fact that Pyrethroids and Pyrethrins are synthetically and toxicologically similar, deterioration when exposed to heat, light, and moistness Gosselin 1984. Pyrethrins with inferiors (Pyrethroids), originally allowed for people Skolarczyk *et al.*, 2017. These pesticides were acquainted due with their lower position of determination when varied with organo chlorine, organophosphate and carbamate fungicides which are related with long haul natural pitfalls. Pyrethroids as a rule break down within the sight of daylight and terrain in a couple of days Bradberry *et al.*, 2005. Also, they do not basically impact ground- water quality making them favored backups to conventional further persisting pesticides.

Pyrethroids' depend on sodium channels and the investiture of delayed depolarization in neurons in the sensory system Silver *et al.*, 2014. WHO believed that pyrethroids are neuro- damages following up on the axons in the supplemental and central nervous system sodium diverts in warm blooded species for by Bradberry *et al.*, 2005 to be multiple times more dangerous to insects than to advanced species. This is because of bugs having further sodium channels, a more modest construction, and lower internal heat position just as lower skin retention and further effective hepatic digestion in warm thoroughbred species. In any case, they've been reckoned for to be dangerous to submarine species e.g., shellfish Lidova *et al.*, 2016, and fin fishes Carcamo *et al.*, 2017, impacting flyspeck directs in both neuronal and mitochondrial layers. They've likewise been shown for to be toxic to advanced invertebrates Ding *et al.*, 2017.

Another manufactured pyrethroid, typically known as Cyphenothrin (5% EC), assessed as a space splash/ mist for its viability against three vector mosquitoes. Cyphenothrin has been reported dangerous by International Program on Chemical Safety (IPCS). Overtures did nearly differently showed issues against mosquito (*Culex pipiens*), housefly (*Musca domestica*), and German cockroach (*Blatella germanica*). Field concentrates in Malaysia with Gokilaht- S 5% EC against Aedes aegypti, *Ae. albopictus* and *Cx. quinquefasciatus* displayed empowering results.

Pyrethroids pose a serious potential hazard to fish because of the use of these compounds in many aquatic larvicidal programs. Pyrethroids are highly toxic to most fish being deltamethrin one of the most toxic and widely used. The fish exhibit several symptoms of stress when treated with cyphenothrin. Emulsifiable concentrate formulations of pyrethroids are usually two to nine times more toxic than the technical grade, most likely due to synergistic interactions Sanchez-Fortun and Barahona, 2005.

Tissue damages brought about by waterborne pollutants can be easily observed because the fish gills come into immediate contact with the environment. The gill surface is more than half of the entire body surface area. In fish the internal environment is separated from the external environment by only a few microns of delicate gill epithelium and thus the branchial function is very sensitive to environmental contamination. Hence, fish serve as excellent bioassay animals for toxicological impact studies and have been widely used for this purpose.

In fish, as in higher vertebrates, the kidney performs an important function related to electrolyte and water balance and the maintenance of a stable internal environment. The kidney excretes nitrogen-containing waste products from the metabolism such as ammonia, urea and creatinin. Following exposure of fish to toxic agents such as insecticide, tissue alterations have been found at the level of the tubular epithelium and glomerulus Teh *et al.*, 1997. Water pollution induces pathological changes in fish. As an indicator of exposure to contaminants, histology represents a useful tool to assess the level of pollution.

Cirrhinus mrigala is an economically important fish and was selected for the bioassay experiments because of its widespread and presently cultured all over the world. They may be easily used in the laboratory and are considered one of the most important non-target organisms in the ponds and streams surrounded by agricultural areas. There is a lack of experimental results about the histopathological effects of cyphenothrin in the literature. The aims of the present study were to study the tissue alterations in the target organs (gill, liver, and kidney) of C. *mrigala* in sub lethal cyphenothrin exposure.

Materials and Methods

Collection of experimental fish

The experiment was carried out in the Environmental Biology and Molecular Toxicological Laboratory, Karnatak University, Dharwad. Healthy *Cirrihinus mrigala,* fishes having a weight range of 100-150 grams were collected from Dharwad local fishponds and lakes. Collected fishes were washed with 0.1% of potassium permanganate ($KMnO_4$) solution to avoid dermal infection. Fishes were then rinsed in water and acclimatized to the laboratory conditions up to 14 days in 25 25 capacity cement tanks. During acclimatization period, fishes were fed with artificial pellet feed and water of the aquaria was also changed once daily. Feeding was stopped 24 h prior to the commencement of the toxicity test experiment.

Experimental procedure

The cyphenothrin toxicity test was carried out in glass aquaria and aquarium contained 20 L of water where six healthy fish were introduced. The definitive toxicity test was carried out using 22. 24. 26, 28, 30µg/L for concentration of cyphenothrin. Each of the toxicant concentration was replicated three times. The LC_{50} of cyphenothrin on the test fish was determined using probit analysis. All data obtained in both tests were analyzed using probit method and the graphical method (Finney, 1971).

Histoarchitecture examination

After the 10, 20, 30- and 40days experiment, gills, liver, Kidney, and specimens were collected from fish alive and were preserved in 10% buffered formalin. Preserved tissues were washed under tap water for 24 h to remove formalin, dehydrated, clarified with xylene, and embedded in paraffin blocks. They were cut at 4-5µ thickness by using Elisa microtome and stained routinely with haematoxylin and eosin (H&E) for histoarchitecture examination. Stained histoarchitecture sections were examined under trinocular microscope. Histoarchitecture changes observed were photographed and interpreted in comparison to the work of previous.

Ethical Statement

All the experiments performed in the present study abide by the guidelines of the Institutional Animal Ethics Committee (IAEC). The experimental animals used in the study were handled with care according to the guidelines provided by the Committee for the Purpose of Control and Supervision of Experiments on Animals (CPCSEA), New Delhi, India.

RESULTS

Fish behavior and mortality

The physical reactions observed in the treated fish were erratic swimming, discolorations of the skin, loss of reflex, hyperactivities, surfacing, and increasing opercula ventilation and these effects increased with increasing concentration of the toxicants and duration of exposure. However, the control group fish (only tap water) was found to be normal behavior during the experimental period. Increased the experimental duration, the treated fish showed an increase in weakness, motion, and gasp for air with slow opercula movement. The 96 h LC_{50} value was found with 30 µg/L but the lower limit (40%) and upper limit (22%), respectively in the experiment.

Histological changes in the Gills of cyphenothrin-treated fish

Fish's gills, which establish direct contact with the medium via which pollutants primarily enter the body, are essential for breathing. There are four pairs of gill arches on teleost fish. The intrabrachial septum is the location of the gill lamellae, which are typically flat, leaf-like structures that have been laterally compressed. The major gill lamellae have a row of secondary gill lamellae on either side of a central rod-like supporting axis. A thin layer of simple squamous epithelial cells is present on the secondary gill lamellae, which have also evolved into respiratory lamellae. These cells are separated by mucus cells and are highly vascularized. Into every secondary gill lamella, blood vessels may be observed extending. One flat, single nucleus can be seen in the blood cell. Interlaminar region refers to the area between two neighboring respiratory lamellae (Fig. 1).

Fish on exposure to the sub lethal concentration of cyphenothrin, on day 10, the enlargement of the base of primary gill lamellae was observed. On day 20 the secondary gill lamellae showed bulging at the distal and basal portions (telangiectatic tissue). Fusion of these lamellae was noticed all along their length. The cells seemed to have undergone a clear increased hyperplasia. On further exposure (day 30) the gills exhibited cytoplasmic vacuolization of superficial epithelial cells, hollow spheres due to greater loss of mucous from mucous cells followed by atrophy of gill lamellae and at day 40 the thinning down of secondary lamellae with desquamation of the hyperplasmic epithelium, loss of the lamellar structure of the gill with cytoplasmic vacuolation and pyknotic nuclei and more number of telangiectatic secondary lamellae were seen in all the gill filaments (Pate 2 Fig.1-6).

In the sublethal concentration of cyphenothrin the gills of the fish exhibited a mild degree of degenerative changes on day 10 of exposure, the slight damage to the base of secondary lamellae and inter lamellar tissue. On day 20 the changes were more marked when compared to day 10. The degeneration of epithelial cells encapsulating primary and secondary gill lamellae with necrosis. Fusion of secondary lamellar, bubbling of primary gill lamellae, atrophy was observed. The excess secretion of mucous and necrosis of basal filament was observed. But on further exposures, for 30 and 40 days the gill structure exhibited mild degree of precipitation of mucus over the gill lamellae (Plate 2 Fig.2-6)

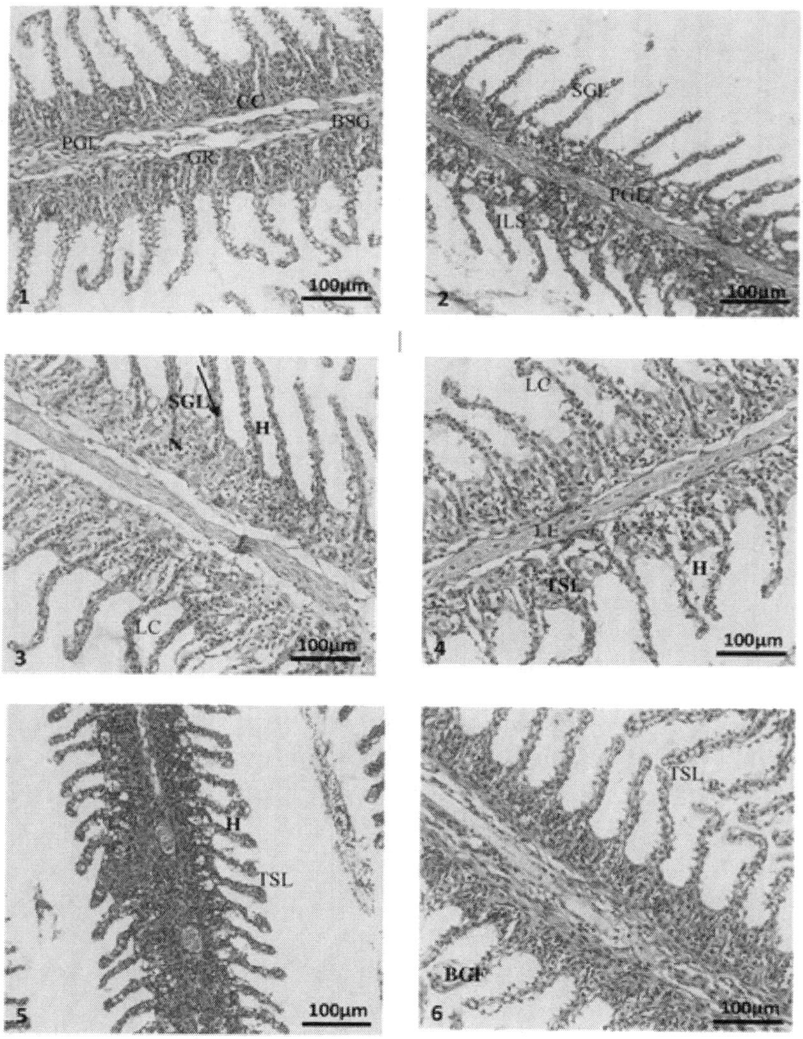

Plate 1–Fig. 1: Section of gill of control fish, *Cirrhinus mrigala* showing normal gill filament (PGL) Primary gill lamellae, (SGL)Intact Secondary gill lamellae, (CC) Chloride Cell, (ILS)Inter lamellar space, (GR) Gill rackers, (BSG)Basal Gill Filament.

Fig. 2: Section of gill of fish, Cirrhinus mrigala exposed to sublethal concentrations of cyphenothrin (EC) (6 µg/L) for day 10 showing loss of secondary gill lamellae and necrosis (N) leading to hyperplasia (HP).

Fig.3-4: Section of gill of fish, Cirrhinus mrigala exposed to sublethal concentrations of cyphenothrin (EC) (6 µg/L) for 20th day showing Telangiectatic secondary lamellae (TSL), lamellar oedema (LE), Hyperplasia (HP).

Fig. 5 and 6: Section of gill of fish, Cirrhinus mrigala exposed to sublethal concentrations of cyphenothrin (EC) (6 µg/L) for 30th and 40th day showing Hyperplasia (HP), and damage of basal gill filament (BGF).

Histological changes in the Liver of cyphenothrin treated fish.

The liver of fish comprises a continuous mass of large hexagonal hepatic cells (Hepatic parenchyma). Hepatic cells are of roundish or polygonal shape containing clear spherical nucleus. They are located among sinusoids forming cord like structures known as hepatic cell cords. In fish these structures are generally obscure. Bile canaliculus is centrally located in each cord. There is no clear division of hepatic cells into lobules. These cells contained granular cytoplasm and with distinct nuclei either exocentric or slightly centrally placed. Hepatic cells have many vital functions other than the secretion of bile. They play an important role in protein, lipid and carbohydrate metabolism. They serve as storage site for some nutrients. Detoxification is another important function. Many blood sinusoids and lipid glycogen granules are found in the hepatic mass (Plate 3. Fig. 1-2).

On day 10 of exposure to the sub lethal concentration of cyphenothrin, the liver exhibited enlarged nuclei and vacuolation in hepatic cells. Liver cords were seen disarrayed. On day 20 of exposure the paranchymatous nature of the liver was greatly disrupted with congested blood vessels. The hapatocytes cell membranes were ruptured and granular degeneration was evident in most of the hapatocytes. Nuclei became slightly hypertrophic. On day 30, severe degrees of atrophic changes were noticed in the liver cords. Hemorrhagic condition was prominent with heavy vacuolization in the liver tissue. At some regions exfoliation and congregation of hepatocytic nuclei and focal necrosis were seen. This was followed by the severe degree of vacuolation, shrinkage of hapatocytes, atrophy, granular degeneration, rupture of blood vessels, necrosis,

Under sublethal concentration of cyphenothrin initially liver exhibited few changes like slight disarray of liver lobes, mild degree of degeneration of cytoplasm, occasional blood clots and congregation of nuclei at day 10 and cloudy swelling of hapatocytes, granulization of cytoplasm, hypertrophic and pyknotic nuclei on day 20. However, on further exposure to day 30 certain degree of reorganization in the structure of liver cords was observed. The nuclei appeared normal, with a very little degree of cytoplasmic vacuolization. At 40 days of exposure, no significant changes were seen, except a slight degree of hyperchromatic condition of the nuclei (Plate 3. Fig.3-6)

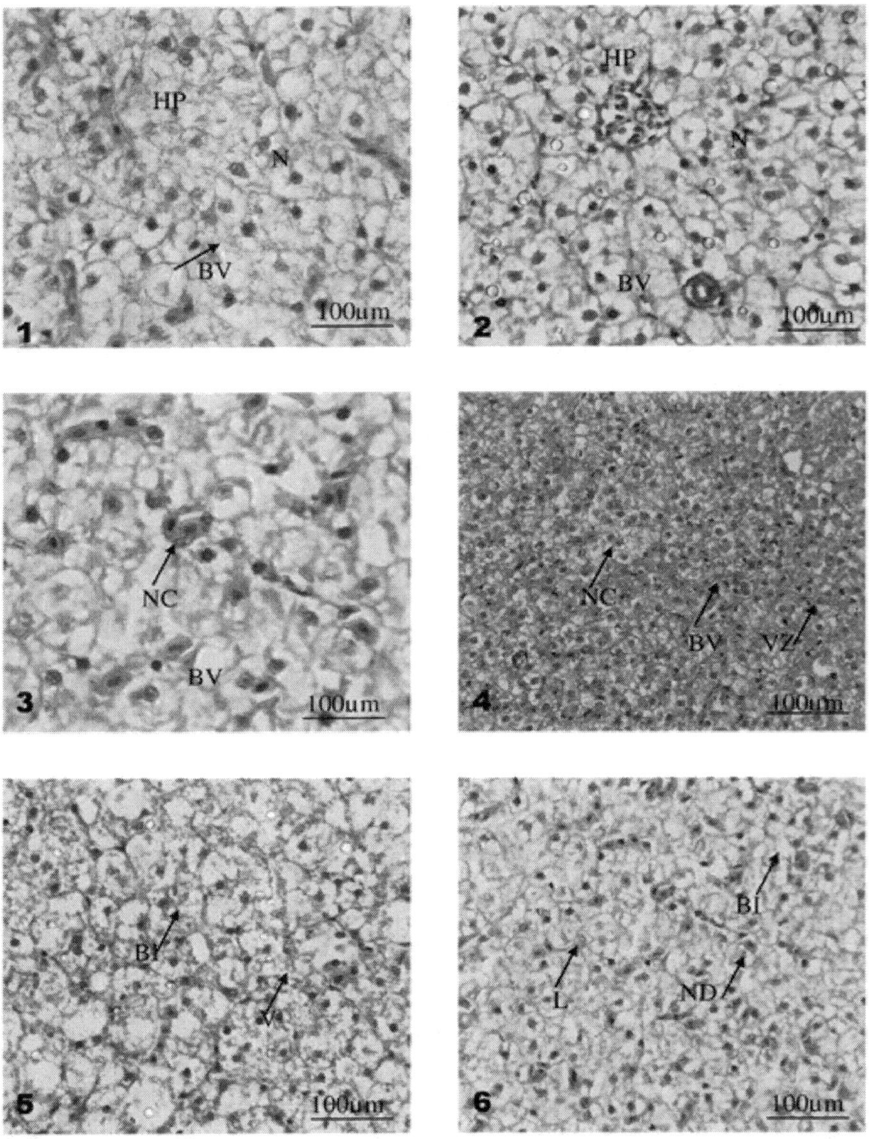

Plate 2. Fig.1: Section of liver of control fish, *Cirrhinus mrigala* showing normal structure (HP) Hepatocytes, (N) Nucleus and (BV) Blood vessel.

Fig.2: Section of liver of control fish, *Cirrhinus mrigala* is a part of liver enlarged showing polygonal hepatocytes (HP) and nucleus (N) and Blood Vessels (BV).

Fig 3: Section of liver of fish, *Cirrhinus mrigala* exposed to cyphenothrin EC (6µg/L) for

10th day showing slight necrosis (NC), damage of blood vessel (BV).

Fig.4: Section of liver of fish, *Cirrhinus mrigala* exposed to cyphenothrin EC (6µg/L) for 20th day showing diffused necrosis (NC), cytoplasmic degeneration, sever damage of blood vessel (BV) and vacuolization of hepatic cells (VZ).

Fig. 5: Section of liver fish, *Cirrhinus mrigala* exposed to cyphenothrin EC 6µg/L) for 30th day showing (BI) blood infiltration; (V) Vacuolation

Fig.6: Section of liver fish, *Cirrhinus mrigala* exposed to cyphenothrin EC 6µg/L) for 40th day showing (L) lipid droplets; (ND) nuclear degeneration, (BI) blood infiltration.

Histological changes in the Kidney of cyphenothrin treated fish

Morphologically, kidney consists of glomerulus, tubules and collecting ducts. The glomerulus is a cluster of capillaries surrounded by the Bowman's capsule. The microscopic picture of the glomerulus shows capillary spaces covered by endothelial cells on the inner side, which in their lumen contain nucleated blood cells. The space between capillaries is filled with mesangial cells. The proximal and distal tubules are covered by tall columnar epithelial cells with oval or round, centrally or basally located nuclei depending on the segment. The cells which cover the tubules are tall and columnar with a weak eosinophilic cytoplasm and showing apical microvilli or brush border towards the lumen (Plate 4; Fig. 1-2).

In sub lethal concentration of cyphenothrin, the kidney showed reduction in renal cell number in the proximal and distal collecting tubules which have resulted in narrowness of lumen. The tubular cells have undergone hypertrophy and some of the renal tubules have lost their normal shape. Vacuolation due to degeneration of cytoplasm is quite obvious. The nuclei of epithelial cells have become quite dominant and are found infiltrating into the surrounding tissue. The perforation of kidney tubules is commonly observed. The kidney demonstrated hyperplasia, vacuolation, degeneration and necrosis leading to the complete necrosis. Cubiodal epithelial cells lining the tubules showed complete vacuolation with degenerating cytoplasm and their disorderly scattering nature. The haemopoitic tissue was fully studded with lymphatic cells at the highest rate of nuclear division. The lumen of the tubules was found to be dilated. Kidney tubules were also found to be perforated (Fig. 1-6).

In sublethal concentration of cyphenothrin the kidney of the fish exhibited a mild degree of changes. On 10th, 20th and 30th day showed more changes i.e., epithelial cells of the tubules which showed desquamation, irregular orientation of the nuclei in the cells, lumen of the tubules became wider as a result of flattening of epithelial cells. Ruptures of the tubules were quite prominent; cell fragments could be seen inside the lumen of some tubules. Vacuolar degeneration was seen in the few tubules. Haemopoitic tissue was degenerated. But on 40th day kidney showed slight recovery tendency. Glomerular cells were disrupted in structure. Cytoplasm appeared with vacuolization and Karyolysis of cell was observed.

Necrotic changes in uriniferous tubules were reduced. Clumping of damaged blood cells was seen (Fig. 1-6).

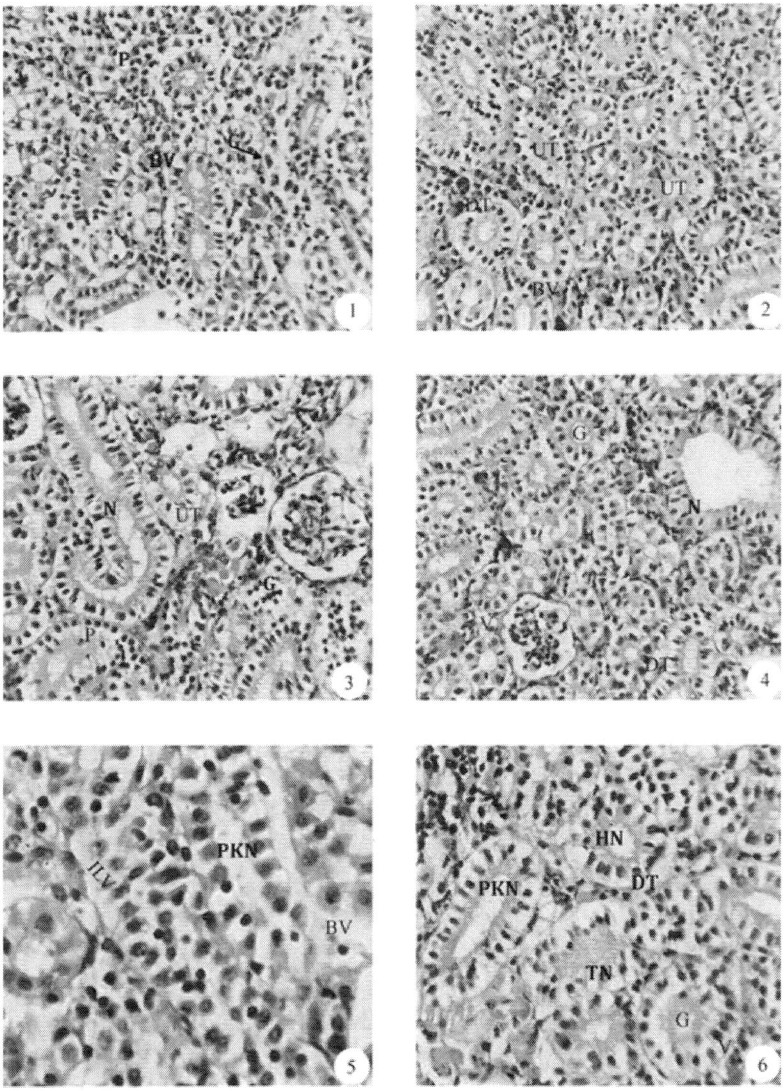

Plate 3; Fig: 1: Section of kidney of control fish, *Cirrhinus mrigala* showing normal structure (P) Proximal tubule; (G)Glomerulus; (BV)Blood vessels; (DT) Distal tubule; (UT) Uriniferous tubule.

Fig: 2: Section of kidney of control fish, *Cirrhinus mrigala* showing an enlarged Proximal tubule (P) with Blood vessels (BV) and Uriniferous tubule (UT).

Fig 3 and 4: Section of kidney of fish, *Cirrhinus mrigala* exposed to sublethal concentrations of cyphenothrin EC (6 µg/L) for 10thand 20thday showing

desquamation and degeneration of tubules, necrosis (N) and Glomerulus shrinkage (G).

Fig 5 and 6: Section of kidney of fish, *Cirrhinus mrigal* **exposed to sub lethal concentration of cyphenothrin EC (6µg/L) for 30th and 40thday showing cellular damages.**

Discussion

A quick way to identify the effects of irritants in various organs is through histopathology Johnson *et al.* 1993. Fish exposed to chemical pollutants are likely to develop a variety of lesions in various organs Bucke *et al.* 1996. To assess the impact of pollution, it is appropriate to histologically examine the gills Poleksic *et al.* 1994, kidney Bucher and Hofer 1993, and liver (ICES 1997).

Aquatic organisms may experience diverse biochemical, physiological, and histological changes in critical tissues because of exposure to lethal and sublethal concentrations of pesticides in their environment Anbu and Ramaswamy 1991; Geraldine *et al.* 1999. In the current study, *Cirrhinus mrigala* exposed to sublethal concentrations of cyphenothrin had histological abnormalities in its gills, liver, and kidney. Dalela *et al.* 1979 reported observing similar results. When rainbow trout were exposed to methiocarb and endosulfan, Altinok and Capkin 2007 discovered a similar outcome.

Due to the transfer of respiratory gases and the maintenance of osmotic and ionic equilibrium, the gills are an essential organ in aquatic life Evans 1993. Toxic agent damage to the gills sets off a series of damaging circumstances that could result in respiratory distress Magare and Patil 2000. Histologically, fish gills exposed to cyphenothrin showed lamellar fusion, lamellar hyperplasia, epithelium hypertrophy and necrosis, separation of the epithelium from the lamellae (epithelial lifting), and eosinophilic material in the area under the epithelium. Sunitha and Sahai 1983, Roy and Munshi 1991, and Bhavan and Geraldine 2000 also made similar discoveries. These research' findings unmistakably show that different effects of sublethal concentrations of cyphenothrin on fish gills exist. Compared to the liver and kidney, the gills were shown to be the most severely impacted organs in the current investigation, possibly because of direct contact with the substance.

Fish exposed to various pesticides have histological abnormalities in their gill tissue, according to several investigators Lowe 1964, Eller 1971, Jauch 1979, Nowak 1992; Rijijohn and Jayabalan 1993. The damage to the epithelial cells and/or hyperplasia, lamellar edema, telangiectasis, and aneurysm were the most common histological abnormalities. The present investigations revealed mild to moderate congestion of the principle lamellae and hyperplasia of branchial plates in the gill tissues. The modifications pointed to a changed oxygen supply to the test fish, producing hypoxic respiratory reactions. Even though it is not fatal, gill damage brought on by environmental toxins is significant in terms of morbidity since it hinders growth and reproduction.

Fish liver histopathology is a valuable tool for examining the consequences of aquatic animal exposure to toxins found in the aquatic environment because it serves as a chemical toxicity indicator Fernandes *et al.* 2008. The liver serves as the main organ for metabolism, xenobiotic detoxification, hazardous material excretion, etc. Individuals with greater levels of pesticides, such as aldrin, alachlor, and dichloroaniline, displayed liver fibrosis, a significant region of necrosis, leukocyte infiltration, and a lack of melanomacrophages Miranda *et al.* 2008. Mathur 1965 demonstrated that the livers of *Ophiocephalus punctatus*, Barbus stigma, *Trichogaster fascitus*, and *Heteropneustes fossilis* all underwent severe liver degeneration as a result of dieldrin. When fish were researched by Hinsen *et al.* in 1971 after being exposed to pesticides in a lab setting, they noted that the liver is the organ having Although there is very little literature on histopathological abnormalities, fish liver histopathology is a helpful technique to examine the consequences of exposure of aquatic animals to toxins the greatest pesticide concentration.

According to Amminikutty and Rege 1977, hepatocytes in liver tissue underwent fast degradation and vacuolation as a result of chronic exposure to thiodon and agallol. According to Mandal and Kulashrestha's research from 1980, *Clarias batrachus* exposed to DDT developed empty blood vessels, necrosis, vacuolation, a breakdown of cell walls, and multinucleated large cells. In their study of the hepatopathology of *Anabas testudineus* from 1984, Ray and Bhattacharya noted changes such as cytoplasm precipitation, cell vacuolation with matching increases in cell size, nuclear degeneration, hepatic cord disruption, etc. Similar to this, John *et al.* 1993 examined the effects of endosulfan on the liver of *Cyprinus carpio* and stated that the corresponding histological changes included widespread vacuolation, fuzzy cell borders, loss of polygonal shape of the cell, and progressive necrosis. As was already noted, our current research on the effects of sublethal doses of cyphenothrin on the liver of *Cirhinus mrigala* during the sub lethal exposure period is significantly corroborated by hepatopathy.

The principal organ for the detoxication of organic xenobiotics is the liver. It suffers negative consequences when a variety of pesticides and other hazardous byproducts tend to build in high concentrations within it. Fish treated to pesticides in a lab setting were analyzed by Hinsen *et al.* 1971, who found that the liver had the greatest pesticide content. Brown trout fry and adult guppies exposed to 0.0032 to 3.2 ppm DDT have many tiny vacuoles in their hepatic cells, according to King 1962. He suggested that this might be connected to the young trout's absorption of the fatty yolk. The degree of structural heterogeneity appears to be a characteristic of inebriated fish livers in general, increasing with the dose of the toxicant Hawkes 1980. When Heteropneustes fossils were exposed to severe thiodan poisoning,

Narayan and Singh 1991 noticed significant cytoplasm degeneration with pyknosis of the nuclei and loss of glycogen in the liver tissue.

The kidney plays a crucial role in the preservation of a stable internal environment as well as the electrolyte and water balance in fish, just like in higher vertebrates. Waste products from metabolism that contain nitrogen, such as ammonia, urea, and creatinine, are expelled by the kidney. Histological changes have been discovered at the level of the tubular epithelium and glomerulus after fish have been exposed to hazardous substances like pesticides Tech *et al.* 1997.

In the current investigation, it was found that kidney exposed to sublethal concentrations of cyphenothrin underwent some significant histological changes. Particularly prominent in the fish kidney were tubular epithelium separations, tubular necrosis, glomeruli shrinkage, renal tubular degeneration, and hyaline droplet degeneration.

Hydrolic changes in the renal tissue may be the cause of the tubular vacuolization, necrosis, and dilatation seen in the kidney after exposure to toxins. This suggests that toxins may cause a failure in the ion pump transport of tubular cells, leading to swelling of the epithelium and tubule degeneration. These changes also imply that renal cells are unable to handle functional disruptions brought on by toxins. According to Sanchez-Chardi *et al.* 2009, tubular necrosis causes the renal excretory function of nephrons to be lost, and dilatation may act as a compensatory mechanism.

It is well known that when exposed to circumstances during reorganization attempts, the renal tissue suffers from substantial structural variation. For instance, a decrease in glomerulus size brought on by variations in water salinity and necrosis on kidney epithelial cells during the spawning season. The animal's homeostasis and health may be harmed by morphologic renal changes since the kidneys play a key role in the development of the immune response Marcelo *et al.* 2000. Following exposure to sublethal levels of cyphenothrin, similar modifications were discovered in the kidney of *Cirrhinus mrigala*.

. One of the first organs to be impacted by water pollutants is the teleostean kidney Thophon *et al.* 2003. The most frequent changes observed in the kidney of fish exposed to contaminated water are tubule degeneration (cloudy swelling and hyaline droplets) and changes in the corpuscle, such as enlargement of glomerulus capillaries and reduction of Bowman's gap Takashima and Hibiya 1995. According to Thophon *et al.* 2003, exposure to metals frequently results in changes to the tubules and glomerulus. Handy and Penrice 1993 discovered swollen Bowman's capsule cells and melanomacrophages in the kidneys of trout *Salmo trutta* and *Tilapia Oreochromis mossambicus* exposed to mercuric chloride. Similar changes were seen in fish exposed to organic pollutants and combined environmental toxins Veiga *et al.* 2002; Schwaiger *et al.* 1997; Pacheco and Santos 2002.

These results imply that the kidney's histological abnormalities, like those in the gills, cannot be attributed solely to stresses. In the current investigation, the fish kidney frequently exhibited hazy edoema in the tubule cells. Cell hypertrophy and the presence of tiny granules, which resemble a net-like structure in the cytoplasm, help to identify this transformation. The degeneration process can develop from this first stage to hyaline degeneration, which is distinguished by the presence of large eosinophilic granules inside the cells.

Conclusions

Thus, the histological changes that were taking place in the present study, as the initial period of exposure in the organs of the fish on exposure to cyphenothrin, toxicity might be a part of defense mechanism. The further accumulation of cyphenothrin in the organs of the fish on prolonged exposure caused destruction in the organ structures. The slight structural reorganization of the gill, liver, and kidney of the fish observed at day 30 of exposure to cyphenothrin toxicity gives support to some extent that the ability of the fish to resist the sublethal stress and in the repair of the damage caused to the organ by enhancing the protein synthetic potentials and other associated activities of the cell.

References

Altinok I and Capkin E. 2007. Histopathology of rainbow trout exposed to sublethal concentrations of methiocarb or endosulfan. Toxicol. Pathol. 35: 405—410.

Amminikutty CK and Rege MS. 1977. Effects of acute and chronic exposure tompesticides, thiodan E.C. 35 and Agallol on the liver of widow tetra Gymnocorymbus temetzi (Boulenger). Indian J. Exp. Biol. 15: 197-200.

Anbu RB and Ramaswamy M.1991. Adaptive changes in respiratory movements of an air breathing fish, Channa striatus (Bleeker) exposed to carbamate pesticide, sevin. J Ecobiol. 3:11-6.

ARJMANDI, R., TAVAKOL, M. AND SHAYEGHI, M., 2010. Determination of organophosphorus insecticide residues in the rice paddies. International Journal Environmental Science Technology, 7(1): 175-182.

BANAEE, M., SUREDA, A., MIRVAGEFEI, A. R. AND AHMADI, K., 2013. Biochemical and histological changes in the liver tissue of Rainbow trout (*Oncorhynchus mykiss*) exposed to sub-lethal concentrations of diazinon. Fish Physiolgy Biochemistery. 39(3): 489- 501.

Bhavan PS and Geraldine P. 2000. Histopathology of the hepatopancreas and gills of the prawn *Macrobrachium malcolmsonii* exposed to endosulfan. Aquat Toxicol. 50:331-9.

Bucher F and Hofer R. 1993. The effects of treated domestic sewage on three organs (gills, kidney, liver) of brown trout (*Salmo trutta*). Wat. Res. 27(2): 255-261.

Bucke D, Vethaak D, Lang T, Mellergaard S. 1996. Common Diseases and Parasites of Fish in the North Atlantic: Training Guide for Identification. International Council for the Exploration of the Sea Techniques in Marine Environmental Sciences, Copenhagen.

Dalela RC, Bhatnagar MC, Tyagi AK, Verma SR. 1979. Histological damage of gills in *Channa gachua* after acute and subacute exposure to endosulfan and rogor.Mikroscopie. 35:301-7.

Eller EE. 1971. Histological lesions in cutthroat Salmo dark exposed chronically to the insecticide endrin. Am. J. Pathol. 64: 321-336.

Evans DH. 1993. The physiology of fishes. Boca Raton, Florida: CRC Press.

Fernandes C, Fontainhas-Femandes A, Rocha E, Salgado MA. 2008. Monitoring pollution in Esmoriz-Paramos lagoon, Portugal: liver histological and biochemical effects in *Liza saliens*. Environ Monit Assess; 145:315-22.

Geraldine P, Bhavan PS, Kaliamurthy J, Zayapragassarazan Z. 1999. Effects of dichlorvos intoxication in the freshwater prawn, *Macrobrachium malcolmsonii*. J Environ Biol. 20:141-8.

Handy RD and Penrice WS. 1993. The influence of high oral doses of mercuric chloride on organ toxicant concentrations and histopathology in rainbow trout, *Oncorhynchus mykiss*. Comparative Biochemistry and Physiology (C), 106: 717- 724.

Hawkes JW. 1980. The effect of xenobiotic on fish tissues: Morphological studies. Fed Proc. 39: 3230-3236.

Hinsen DJ, Parrish PR, Loser JI, Wilson AJ, Wilson PD. 1971. Chronic toxicity uptake and retention of arochlor 1254 in two estuarine fishes. Bull. Environ. Contam. Toxicol. 6: 113-119.

ICES (International Council for the Exploration of the Sea). 1997. Report of the Special Meeting on the Use of Liver Pathology of Flat fish for Monitoring Biological Effects of Contaminants. International Council for the Exploration of the Sea, ICES CM, 1997/F: 2,pp75.

J. Lidova,A. Stara,A. Kouba, J. Velisek. 2016. The effects of cypermethrin on oxidative stress and antioxidant biomarkers in marbled crayfish (*Procambarus fallax f. virginalis*) , Neuro Endocrinol. Lett. 37: 53–59.

J. Skolarczyk, J. Pekar, B. Nieradko-Iwanicka, 2017. Immune disorders induced by exposure to pyrethroid insecticides, Postepy Hig. Med. 71 446–453.

J.G. Cárcamo, M.N. Aguila, C.F. Carreño, T. Vera, L. Arias-Darraz, J.E. Figueroa, A.P. Romero, M. Alvarez, A.J. Yanez. 2017. Consecutive emamectin benzoate and deltamethrin treatments affect the expressions and activities of detoxification enzymes in the rainbow trout (Oncorhynchus mykiss), Comp. Biochem. Physiol. C Toxicol. Pharmacol. 191 129–137.

Jauch D. 1979. Gill lesions in Cichlid fishes after intoxication with the insecticides Fenthion. Experimentia, 35:371-372.

John KR, Jayabalan N, George MR. 1993. Impact of sublethal concentrations of endosulfan on the histology of Cyprinus carpio liver and kidney. In: Proceedings of the National Seminar on Aquaculture Development in India Problems and Prospects, pp. 179-182.

Johnson LL, Stehr CM, Olson OP, Myers MS, Pierce SM, Wigren CA, McCain BB, Varanasi U. 1993. Chemical contaminants and hepatic lesions in winter flounder (Pleuronectes americanus) from the Northeast Coast of the United States. Environ Sci Technol. 27: 2759-71.

K.S. Silver, Y. Du, Y. Nomura, E.E. Oliveira, V.L. Salgado, B.S. Zhorov, K. Dong. 2014. Voltage-Gated Sodium Channels as Insecticide Targets, Advancement in Insect Physiology 46: 389–433.

King SF. 1962. Some effects of DDT on the guppy and the brown trout. U.S. Fish Wildlife Serv. Spec. Sci. Res. Fish. 20: 399.

L.G. Costa. 2015. The neurotoxicity of organochlorine and pyrethroid pesticides, Handbook. Clin. Neurol. 131 135–148.

Lowe JI. 1964. Chronic exposure to *Leiostomus xanthurus* L. to sublethal concentration of toxophene in sea water. Trans. Amer. Fish. Soc. 93: 396-399.

Magare SR and Patil HT. 2000. Effect of pesticides on oxygen consumption, red blood cell count and metabolites of a fish, *Puntius ticto*. Environ Ecol. 18: 891- 4.

MAHBOOB, S., NIAZI, F., ALGHANIM, K., SULTANA, S., AL-MISNED, F. AND AHMED, Z., 2015. Health risks associated with pesticide residues in water, sediments and the muscle tissues of *Catla catla* at Head Balloki on the River Ravi. Environmental Monitoring Assessment, 187: 81

Mandal PK and Kulashrestha H. 1980. Histopathological changes induced by the sublethal sumithion in *Clarias batrachus* (Linn.). Indian J. Exp. Biol. 18: 547-552.

Marcelo LdV, Edson dLR, Fabio JP, Maria JTRP 2000. Histopathologic Changes in the Kidney Tissue of *Prochilodus lineatus* Valenciennes, 1836 (Characiformes, Prochilodontidae) Induced by Sublethal Concentration of Trichlorfon Exposure. Brazilian Archives of Biology and Technology.45 (2): 171-175.

Mathur DS. 1965. Histological changes in the livers of certain fishes induced by dieldrin. Sci. Cult. 31: 258-259.

Miranda AL, Roche H, Randi MAF, Menezes ML, Oliveira Ribeiro CA. 2008. Bioaccumulation of chlorinated pesticides and PCBs in the tropical freshwater fish *Hoplias malabaricus*: histopathological, physiological, and immunological findings. Environ. Int. 34: 939-949.

Narayan AS and Singh BB. 1991. Histopathological lesions in *Heteropneustes fossilis* subject to acute thiodan toxicity. Acta Hydrochem. Hydrobiol. 19: 235-243.

Nowak B. 1992. Histological changes in gills induced by residues of endosulfan. Aquat. Toxicol. 23: 65-83.

P.K. Gupta. 2018. Illustrated Toxicology: With Study Questions,Academic Press.

Pacheco M and Santos MA. 2002. Biotransformation, genotoxic and histopathological effects of environmental contaminants in European eel (*Anguilla anguilla* L.). Ecotoxicology and Environmental Safety: 53: 331-347.

Poleksic V and Mitrovic-Tutundzic V. 1994. Fish gills as a monitor of sublethal and chronic effects of pollution. In: Muller and R., Lloyd, R. (Eds.), Sublethal and Chronic Effects of Pollutants on Fresh water Fish. FAO, Fishing News Books, Oxford, UK, pp 339-352.

R. Ding, Z. Cao, Y. Wang, X. Gao, H. Luo, C. Zhang, S. Ma, H. Jin, C. Lu. 2017. The implication of p66shc in oxidative stress induced by deltamethrin, Chem. Biol. Interact. (278)162–169.

R.E. Gosselin. 1984. Clinical Toxicology and Commercial Products, Williams and

RAHIMINEZHHAD, M.; SHAHTAHERI, S.J.; GANJALIM, R.; RAHIMIFOROUSHANI, A.G.F., 2009: Moleculary imprinted soild phase extraction for trace analysis of diazinon in drinking water. Iranian Journal of Environmental Health Science and Engineering 6: 97-106.

Ray AK, Bhattacharya S. 1984. Histopathological changes in the hepatopancreas of the fresh water air breathing teleost *Anabas testudineus* (Bloch) exposed to acute and chronic levels of cythion. J. Curr. Biosci. 1:170-174.

Rijijohn K and Jayabalan N. 1993. Sublethal effect of endosulfan on the histology and protein pattern of *Cyprinus carpio* gill. J. Appl. Ichtyology. 9:49-56.

Roy PK and Munshi, S. 1991. Malathion induced structural and morphometric changes of gills of a freshwater major carp *Cirrhinus mrigala* (Ham). J. Environ. Biol. 12:79-87.

S.M. Bradberry, S.A. Cage, A.T. Proudfoot, J. Vale. 2005. A Poisoning due to pyrethroids, Toxicology Review 24 (2) 93 – 106.

CHAPTER 2

CYTOTOXIC ACTIVITY OF ZNO NPS SYNTHESIZED FROM AQUEOUS EXTRACT OF *OLDENLANDIA CORYMBOSA* L. INHIBITS CELL PROLIFERATION AND INDUCES CELL APOPTOSIS IN HUMAN BREST CANCER CELL LINES (T-47D)

Anusha M[1], Sushmitha C[1], Shivanand Bhat[2], Siddappa B. Kakkalameli[1]

Department of Studies in Botany, Davangere University, Davangere, Karnataka India.
Department of Botany Smt. Indira Gandhi Government First Grade Womens College, Sagara-577401
***Corresponding author:** *siddubotdu@davangereuniversity.ac.in*

Abstract:

In the present study, zinc oxide nanoparticles (ZnO NPs) synthesized by simple and eco-friendly methods using *Oldenlandia corymbosa* (OC) extract were studied *in vitro*. The characterization of ZnO NPs was done by UV-visible spectroscopy, Fourier transform infrared (FTIR) spectroscopy, Energy dispersive X-ray (EDX) analysis and Scanning electron microscope (SEM) analysis. The formation of ZnO NPs and reduction of zinc ions was observed by changing its colour from light yellow to dark brown within 15 minutes. The UV-visible spectroscopy showed a maxima absorption peak at 400 nm indicating mainly its formation. The FTIR analysis was monitored to identify functional groups and biomolecules in the OC extract as responsible for the capping and stabilization of ZnO NPs. The EDS analysis was used to determine the elemental composition of nanoparticles. The crystalline size of the ZnONPs was confirmed by X-ray diffraction analysis, which was found to be 41.02 nm. The surface morphology and size of the ZnONPs were determined using SEM analysis. The prepared ZnO NPs

Keywords: *Oldenlandia corymbosa* L.; Green synthesis; Characterization; Anticancer activities.

Introduction

Currently, nanotechnology has accomplished more attention in the various domains of nanoscience and technology due to their uniqueness in medicine, bio-labelling, molecular sensing, optical imaging, electronics, food industry and agriculture (Ansari and Alzohairy, 2018). Nanotechnology is defined as the fabrication, characterization and exploration of materials in the range between 1 to 100nm (Ying et al. 2022). These nanoscale materials are used extensively in such areas as biomedicine, engineering, agriculture and biology (Zhao et al. 2016). To provide novel nanoparticles a variety of techniques are applied in different domains such as physics, chemistry and biology. The biological methods have additional advantages over the physical and chemical methods. The physical and chemical routes are often expensive and involve evaporation, radiations, condensation, high temperature, lethal chemicals and reagents during their synthesis (). Whereas, biological methods are stable, viable, safe to design, inexpensive, prolong life span, nontoxic and easy to implement (). Thus, in biological systems algae, bacteria, fungi, enzymes and plant-mediated extracts are employed for the fabrication of nanoparticles (Singh et al. 2018; Daphedar et al. 2020). For example, biological materials acted as capping, reducing and stabilizing agents of synthesized NPs (Fadoun et al. 2021).

Bio-mediated synthesis of nanoparticles including silver (Ag), zinc (Zn), gold (Au), magnesium sulphate ($MgSO_4$), copper sulphate ($CuSO_4$), and titanium dioxide (TiO_2) are easy to synthesize and possess high potential biological activities (Mohd Yusof et al. 2019). Zn and ZnO nanoparticles are safe for humans and animals due to their biocompatible properties (). Recent studies reported that ZnO NPs exhibit prominent antimicrobial and antioxidant activities owing to their tiny size, anisotropic shape and broad surface reactivity (Ali et al. 2018). The ZnO NPs are extremely lethal to the most of the microorganisms.

Oldenlandia corymbosa L. (*syn. Hedyotis corymbosa (L).* Lam) belongs to the family Rubiaceae commonly known as flat-top mille grains (Walter and Lewis, 1964). It is an annual herb mainly growing in monsoon in tropical regions of Asia (Patel et al. 2014). The extracts possess anticancer, antiulcer, antioxidant, hepatoprotecant, antimicrobial, analgesic, and antimalarial activities that have been reported (Noiarsa et al. 2008). The whole plant is used as a decoction to treat various life-threatening chronic diseases, as reported by Archana et al. (2020). Thus the proposed investigation was undertaken to study the green synthesis, characterization and cytotoxic activity of ZnO NPs synthesized by OC on T-47D cancer cell lines. Hence the plant-mediated extract of OC-fabricated ZnONPs could be used as modern medicine in clinical development.

Materials and methods

Test chemicals

The test chemicals and regents like zinc nitrate ($ZnNO_3$), Hcl, NaOH, potassium bromide (KBr) etc were procured from Hi-media lab in Mumbai, India.

Plant material collection

For the experiment, the *Oldenlandia corymbosa* (OC) the whole plant was collected from the campus of Davangere University, Shivagangotri Davangere, Karnataka, India.

Preparation of extract

The collected plant was washed with running tap water followed by double distilled water to remove unwanted dust particles from the plant surfaces. After washing, the plant material was shade-dried at room temperature for 24 hrs. About 15g of dried plant material was incised into small pieces and added into 250 ml of Erlenmeyer flask containing 100 ml of distilled water. Thereafter, aqueous extract was boiled for 1 hrs at 60°C and the mixture was allowed to cool at room temperature. Then, the suspension was filtered through Whatman No.1 filter paper and the filtrate was preserved at 4°C in at refrigerator for further analysis.

Phytochemical analysis

Test for alkaloids:

Dragendoff's test: For about 1ml of the plant extract was taken and added to 1ml of hydrochloric acid (HCl). The mixture was gently mixed with a glass rod and 2-3 drops of Dragendoff's reagent, a cream colour precipitate formation indicates the presence of alkaloids.

Test for Tannins

Braymer's testing: Around 1ml of plant extract was mixed with 2ml of distilled water and heated in a water bath for 10 minutes. The mixture was filtered and ferric chloride was added to the filtrate, a blackish-blue colour indicates the presence of tannins.

Test for steroids:

Lieberman Burchard tests: About 0.2ml of plant extract was mixed with 0.2ml of concentrated sulphuric acid. The formation of a green colour indicates the presence of steroids.

Test for saponins:

Frothing test: Around 0.2ml of plant extract was added to 0.6ml of distilled water in a test tube. The mixture was vortexed and the formation of honeycomb froth indicates the presence of saponins.

Test for flavonoids:

Alkaline reagent test: About 1ml of a plant extract containing a few drops of sodium hydroxide. The formation of a deep yellow colour indicates the presence of flavonoids.

Test for carbohydrates

Benedict's test: About 2ml of plant extract was mixed with a few drops of Benedict's reagent. The mixture is gently heated; the appearance of an orange-red precipitate indicates the presence of carbohydrates.

Green synthesis of ZnO NPs

For green synthesis, we used 10ml aqueous plant extract and added to 90ml of zinc nitrite ($ZnNO_3$) solution. Further, the reaction mixture was incubated in a water bath for 15 minutes at 60°C. After that, the reaction mixture was adjusted to different pH like pH 8, 9 and 10 (with 0.1N HCl and 0.1N NaoH). The colour change was observed from light yellow to dark brown indicating the reduction of zinc ions and the formation of zinc oxide nanoparticles (ZnO-NPs).

Characterization

The absorption spectrum of ZnO NPs synthesized using OC was measured using UV-visible spectroscopy (UV-1650PC-Shimadzu Europe) in the wavelength range between 300 to 800 nm with a resolution of 1nm and distilled water as blank. The purified sample was oven-dried overnight at 45°C. The obtained crystalline powder is mixed with potassium bromide (KBr) to get pellet form. The samples were analyzed in FTIR spectroscopy with an absorption wavelength range between 4000–400 cm^{-1} at resolution 4cm^{-1}. The presence of phytochemicals in synthesized ZnO NPs was responsible for the bio-reduction, capping and stabilization of nanoparticles was investigated. Further Energy dispersive X-ray analysis was used to validate the elemental composition of the fabricated nanoparticles by

INSPECT F50 (FEI, Netherlands). The sample purity and the size of the ZnONPs were determined by X-ray diffraction analysis. The morphological evaluations of ZnONPs were monitored by a scanning electron microscope (SEM) analysis.

Cell viability test

In this study, the given test compound is evaluated to analyse the cell cytotoxicity on T-47D cell lines and blank as media without cells. Cancer cells were taken into 4-5 plates 3 cells in each plate were allowed to different concentrations of ZnONPs viz. 31.25, 62.5, 125, 250, 500µG/ml were taken to treat the cells for 48 hours and standard as T-47D treated with 12.5 µM quinacrine. Thereafter MTT colour is added to all the concentrations and colour changes were observed through a spectrophotometer. The readings were recorded with different OD detects the concentration; depending on colour change viability is predicted. By this process, cell death is confirmed and DNA degradation can be identified. The concentrations of the test compound used to treat the cells are as follows:

Table 1

Sl.No	Test Compounds	Cell Line	Concentration treated to cells
1	Untreated	T-47D	No treatment
2	Standard (C)	T-47D	12.5 µM quinacrine.
3	Blank	-	Only Media without cells
4	OC(ZN-3)	T-47D	5(31.25,62.5,125,250,500µG/ml)

Results and discussion

UV-Visible spectroscopy analysis

The whole plant extract was used to evaluate the optical properties of synthesized ZnONPs from UV-visible spectra analysis. The 7.5ml of plant extract was added to 250ml of Erlenmeyer flask containing 2 mM of zinc oxide. The reaction mixture was heated at 65°C for 20 minutes and the sample was allowed to cool at room temperature. The colour of the reaction mixture changed from pale peach to maroon to dark brown colour within 15 minutes indicating the reduction of metal ions and the formation of ZnO nanoparticles (Jayachandran et al. 2021). Thereafter, the reaction mixture was adjusted to different pH such as pH 8, 9 and 10. Further, the sample was analyzed in UV-visible spectrum and distilled water as blank. The spectra of the ZnO-NPs showed a maximum absorption peak at 400 nm, corresponding to a band gap of 4.5 eV at pH 8. Whereas, pH 9 and pH 10 show broad absorption peaks at 300 and 500 nm wavelengths with the band gap found to be 3.4 and 4.8 eV respectively (Sasani et al. 2016). This pattern of absorption spectrum can be assigned to the intrinsic band gap absorption of ZnO-

NPs due to the electron transition from the valence band to the conduction band (Udayabhanu et al 2015). The band gap enhancement arises due to the size effect of the NPs. The sharp peak indicates that the particles are nano in size, and the particle size distribution is narrow.

Fig 1: Formation of brown colour indicates Zno nanoparticles

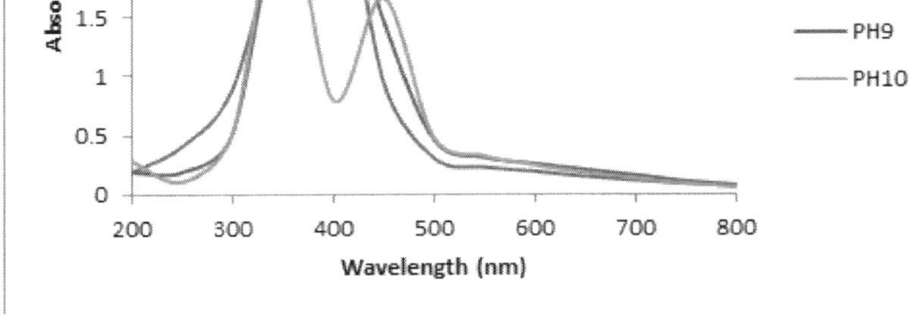

Fig 2: UV-visible analysis of ZnO NPs synthesized from OC plant extract

FTIR ANALYSIS

FTIR analysis was carried out to analyze biological moieties present in the ZnONPs. The spectrum shows characteristic absorption peaks at 3349, 2360, 2964, 1625, 1700, 1089 and 628cm^{-1}. The prominent bands 3349cm^{-1} correspond to the hydroxyl group stretching of phenol and alcoholic group (Umamaheswari et al.

2021). The sharp intense band 2360cm^{-1} is assigned due to C=C conjugated (or antisymmetric stretching mode of CO_2). The band 2964cm^{-1} will show an extremely prominent peak which corresponds to the CH and CH aliphatic stretching group. The peak of 1625 is attributed to the OH bending which condenses moisture from the sample (Upadhyay et al. 2020). The weak band 1700cm-1 was assigned due to C=O, C-O and C-H vibrations (Sowri Babu et al. 2013). The peak 1089cm^{-1} arises due to the formation of phospholipids (P=O). The peak 540cm-1 described the bonding of zinc oxide which confirms the presence of ZnO particles (Geetha et al. 2016)

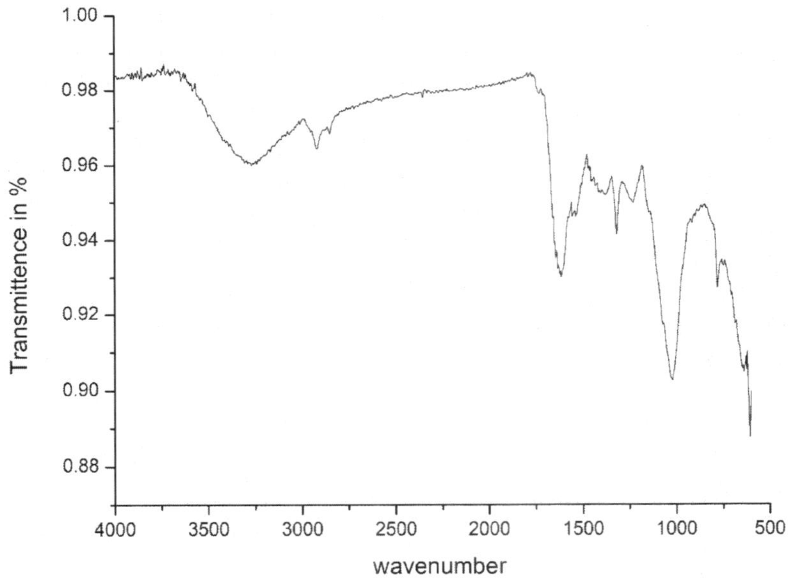

Fig 3. FTIR analysis of ZnO nanoparticles synthesized by plant extract of OC

Table 2. FTIR analysis and their functional groups present in ZnO nanoparticles

Sl.NO.	Absorption peaks (cm^{-1})	Functional groups
1	3349	Hydroxyl (OH) group stretching vibration of phenol or alcoholic group
2	2360	C=C conjugated (or antisymmetric stretching mode of CO_2)
3	2964	CH and CH aliphatic stretching group
4	1625	OH bending
5	1700	C=O, C–O and C–H vibrations
6	1089	P=O phospholipids in DNA and RNA
7	540	ZnO bonding

Phytochemical analysis

The results of the phytochemical analysis of plant extract are supported by the results of FTIR. Phytochemical compounds like alkaloids, flavonoids, steroids, saponins, carbohydrates, starch and tannins are present in the plant extract. These compounds are involved in the reduction, capping and stabilization of nanoparticles. The presence of functional groups like hydroxyl, carboxyl and aliphatic stretching groups shows phytochemicals present in the plant extract. These functional groups reduce zinc ions and are involved in the formation of zinc oxide nanoparticles.

EDX analysis

EDX analysis was carried out to confirm Zn in the oxide form. The formation of peaks exhibits the elemental composition of the ZnO nanoparticles (Nagarajan and Kuppaswamy et al 2013). The EDX spectrum of ZnO-NPs gives a peak around 1.0 Kev, which belongs to the binding energy of Zn and other peaks located at binding energies of 0.4, 0.5, and 2.0 correspond to Zn. The obtained product was pure since there were no impurity peaks in the spectra. These values which confirm the elemental composition of the compound synthesized from the plant extract of OC.

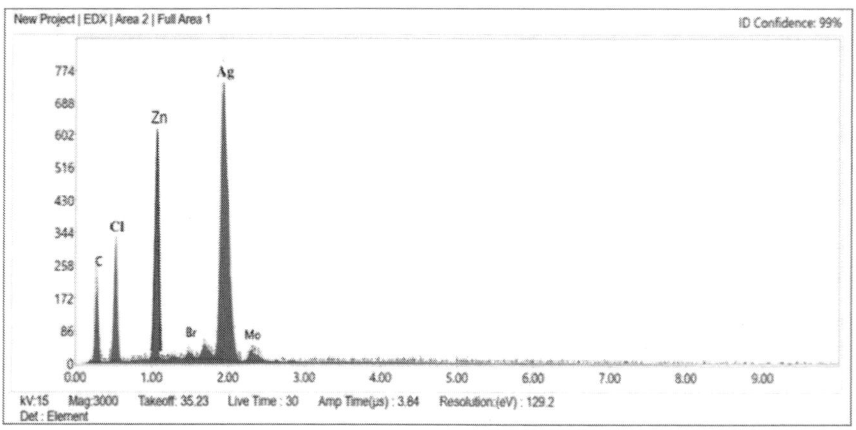

Fig. 4. EDX analysis of synthesized ZnONPs from the plant extract of OC.

XRD analysis

The crystalline purity and size of the ZnONPs synthesized from OC plant extract were confirmed by X-ray diffractometer analysis (Fig. 5). The crystalline powder scanning was done in the 2θ angle range between 20° to 80°. The highest peak (111) represents the purity and structure of nanoparticles. The size of the ZnONPs was calculated using Debye-Scherrer's equation and the size was found to be 41.02 nm. This result was compared with the pattern of standard XRD (JCPDS file no. 36-1451). Similar results have been observed in ZnO NPs synthesized using *Myristica fragrans* fruit extracts (Faisal et al. 2021).

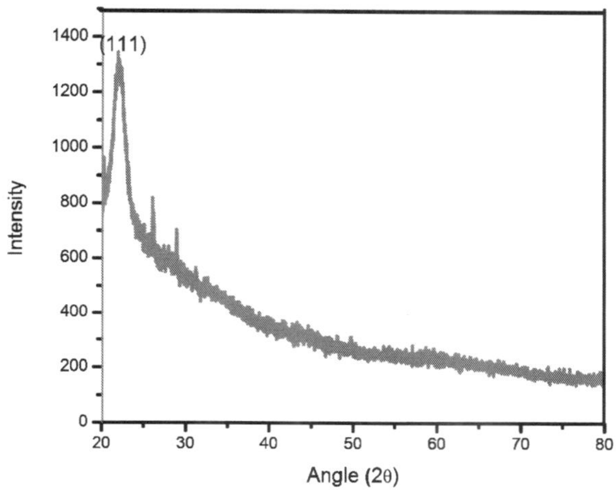

Fig. 5: The crystalline purity and size of the ZnONPs synthesized from Ocimum plant extract by X-ray diffractometer analysis

SEM analysis

SEM images were used to explore the surface morphology, size and aggregation of ZnO-NP. The SEM micrographs showed that morphology comprises voids and pores, the cause of which could be treated to the large amounts of hot gasses that escape out of the reaction mixture during a combustion process. Through the pores of various sizes and shapes the crystallites are interlinked to one another, it is significant that the agglomerated crystallites are uniformly spherical.

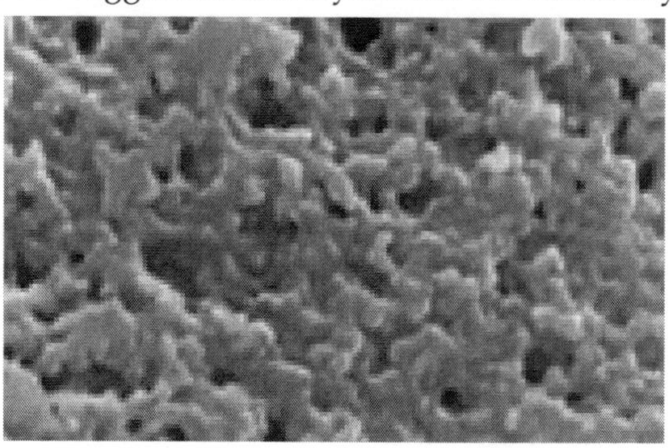

Fig 6: SEM micrograph of ZnO-NPs

Anticancer activity

In this study, the given test compound is evaluated to analyze the cell cyto toxicity on T-47D cell lines. The concentrations of the test compound used to treat the cells are as follows:

Table 3: Anticancer activity analysis of ZnO-NPs

Sl.No	Test Compounds	Cell Line	Concentration treated to cells
1	Untreated	T-47D	No treatment
2	Standard (C)	T-47D	12.5 µM quinacrine.
3	Blank	-	Only Media without cells
4	OC(ZN-3)	T-47D	5(31.25,62.5,125,250,500µG/ml)

Concentration Unit: µg/ml		Incubation: 48hrs						
Concentration	Blank	**untreated**	STD	31.25	62.5	125	250	500
Cell Viability %		100	3.167155	98.26979	**88.56305**	80.58651	69.50147	**21.70088**
IC 50 VALUE=334.1 µg/ml								

Cancer cells were taken into 4-5 plates 3 cells in each plate they are allowed to different concentration like standard it may be positive are negative and then 31.25, 62.5, 125, 250, 500µG/ml concentrations were taken to treat the cells. After 48 hours, MTT colour is added to the all concentrations, later colours changes were observed through spectrophotometer. Moreover, its readings with different OD detects the concentration, depending on colour change viability is predicted. By this process cell death is checked and DNA degradation can be identified. Where, in standard concentration about only 3.1671 55% of cells were died, 31.25 concentration 1.73021%, 62.5 concentration 11.43695%, 125 concentration 19.41349, 250 concentration 30.49853%, 500 concentration 78.29912% of cells were died. Hence, it concludes that increase in concentration leads to large numbers of cells death.

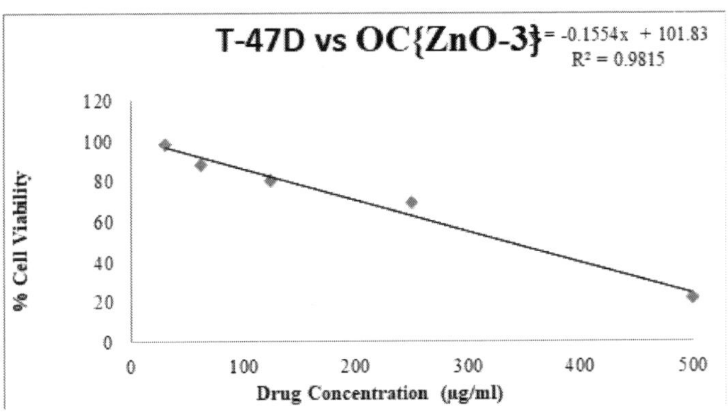

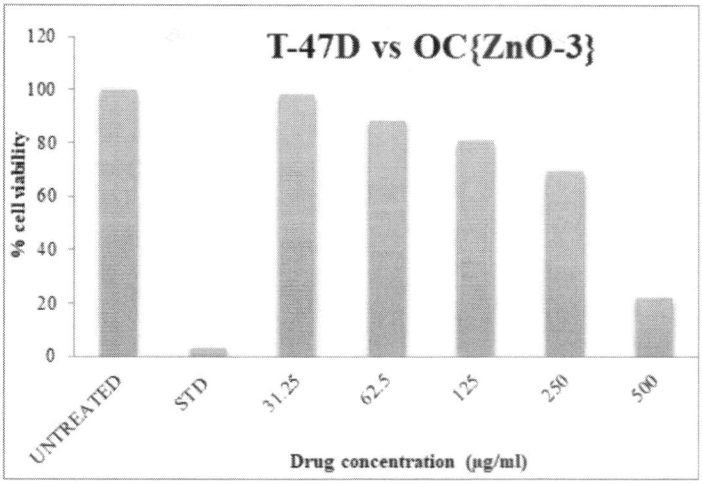

Fig. 7: Anticancer activity analysis of ZnO-NPs

By the graph it is clear that in the concentration of drug results in increase in number of cell deaths and cell viability can be predicted by the drug concentrations. Hence, it concludes that increase in drug concentration leads to more number of cell deaths.

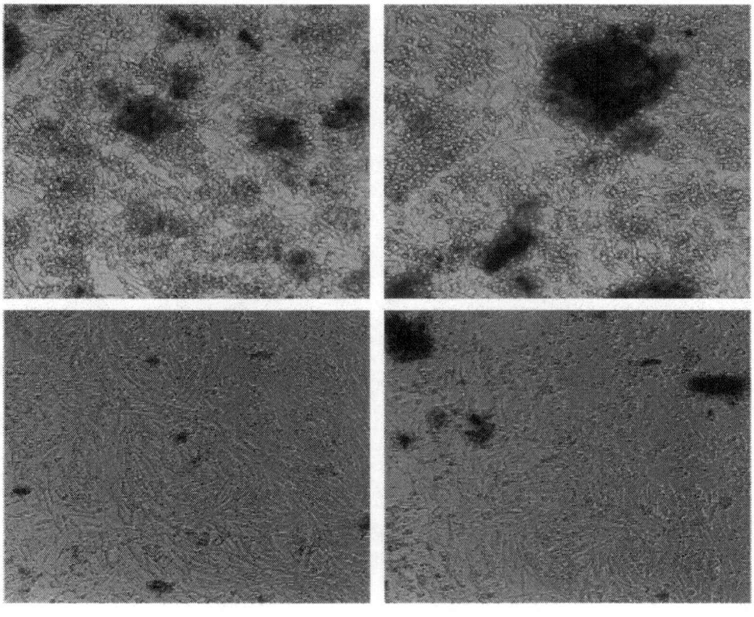

Fig. 8: Images of anticancer activity

References

Ali A, Phull A and Zia M. 2018. Elemental zinc to zinc nanoparticles: is ZnO NPs crucial for life? Synthesis, toxicological, and environmental concerns. Nanotechnol Rev. 7(5): 413–441.

Ansari MA and Alzohairy MA. 2018. One-Pot Facile Green Synthesis of Silver Nanoparticles Using Seed Extract of *Phoenix dactylifera* and Their Bactericidal Potential against MRSA. Evidence-Based Complementary and Alternative Medicine. Article ID 1860280.

Archana V., Thomas N. N. , Lakshmi S., Rauf A.A., Edwin B.T. (2020). Pharmaceutical properties of *Oldenlandia corymbosa* Linn. Materials today: proceedings. , https://doi.org/10.1016/j.matpr.2020.05.585.

Ashwini Jayachandran *, Aswathy T.R., Achuthsankar S. Nair. Green synthesis and characterization of zinc oxide nanoparticles using *Cayratia pedata* leaf extract. Biochemistry and Biophysics Reports 26 (2021) 100995.

Faisal S, Jan H, Shah SA, Shah S, Khan A, Akbar MT, Rizwan M, Jan F, Wajidullah, Akhtar N, Khattak A, Syed S. Green Synthesis of Zinc Oxide (ZnO) Nanoparticles Using Aqueous Fruit Extracts of *Myristica fragrans*: Their Characterizations and Biological and Environmental Applications. ACS Omega. 2021 Mar 30;6(14):9709-9722.

Jadoun, S., Arif, R., Jangid, N.K. *et al.* Green synthesis of nanoparticles using plant extracts: a review. *Environ Chem Lett* **19**, 355–374 (2021).

K. Sowri Babu, A. Ramachandra Reddy, Ch. Sujatha, K. Venugopal Reddy, A. N. Mallika. 2013. Synthesis and optical characterization of porous ZnO. Journal of Advanced Ceramics. 2(3): 260–265.

M. Sasani Ghamsari, S. Alamdari, W. Han, H. Park, Impact of nanostructured thin ZnO film in ultraviolet protection, Int. J. Nanomed. 12: (2016) 207–216.

M.S. Geetha, H. Nagabhushana, H.N. Shivananjaiah. 2016. Green mediated synthesis and characterization of ZnO nanoparticles using *Euphorbia Jatropa* latex as reducing agent. Journal of Science: Advanced Materials and Devices. 1: 301-310.

Mohd Yusof, H., Mohamad, R., Zaidan, U.H. *et al.* Microbial synthesis of zinc oxide nanoparticles and their potential application as an antimicrobial agent and a feed supplement in animal industry: a review. *J Animal Sci Biotechnol* **10**, 57 (2019).

Noiarsa P, Ruchirawat S, Otsuka H, Kanchanapoom T. (2008) Chemical constituents from *Oldenlandia corymbosa* L. of Thai origin. J Nat Med. 62:249–250.

Patel T. Jain V. Dodia R. (2014) *Oldenlandia corymbosa* L.: A Phytopharmacological review. International Journal of Phytopharmacy. 4(3), 79-82.

Singh, J., Dutta, T., Kim, KH. *et al.* 'Green' synthesis of metals and their oxide nanoparticles: applications for environmental remediation. *J Nanobiotechnol* **16**, 84 (2018)

Upadhyay P K, Vikas Kumar Jain, A K Shrivastav, Shashank Sharma and Ravi Sharma. Green and chemically synthesized ZnO nanoparticles: A comparative study. Materials Science and Engineering 798 (2020) 012025. doi:10.1088/1757-899X/798/1/012025.

Walter H. Lewis (1964) Oldenlandia Corymbosa (Rubiaceae), Grana Palynologica, 5:3, 330-341,

Ying S, Guan Z, Ofoegbu PC, Clubb P, Rico C, He F, Hong J. 2022. Green synthesis of nanoparticles: Current developments and limitations. Environmental Technology & Innovation. 26, 102336.

Zhao X, Liu W, Cai Z, Han B, Qian T, Zhao D. 2016. An overview of preparation and applications of stabilized zero-valent iron nanoparticles for soil and groundwater remediation. Water Research. 100 (1), 245-266.

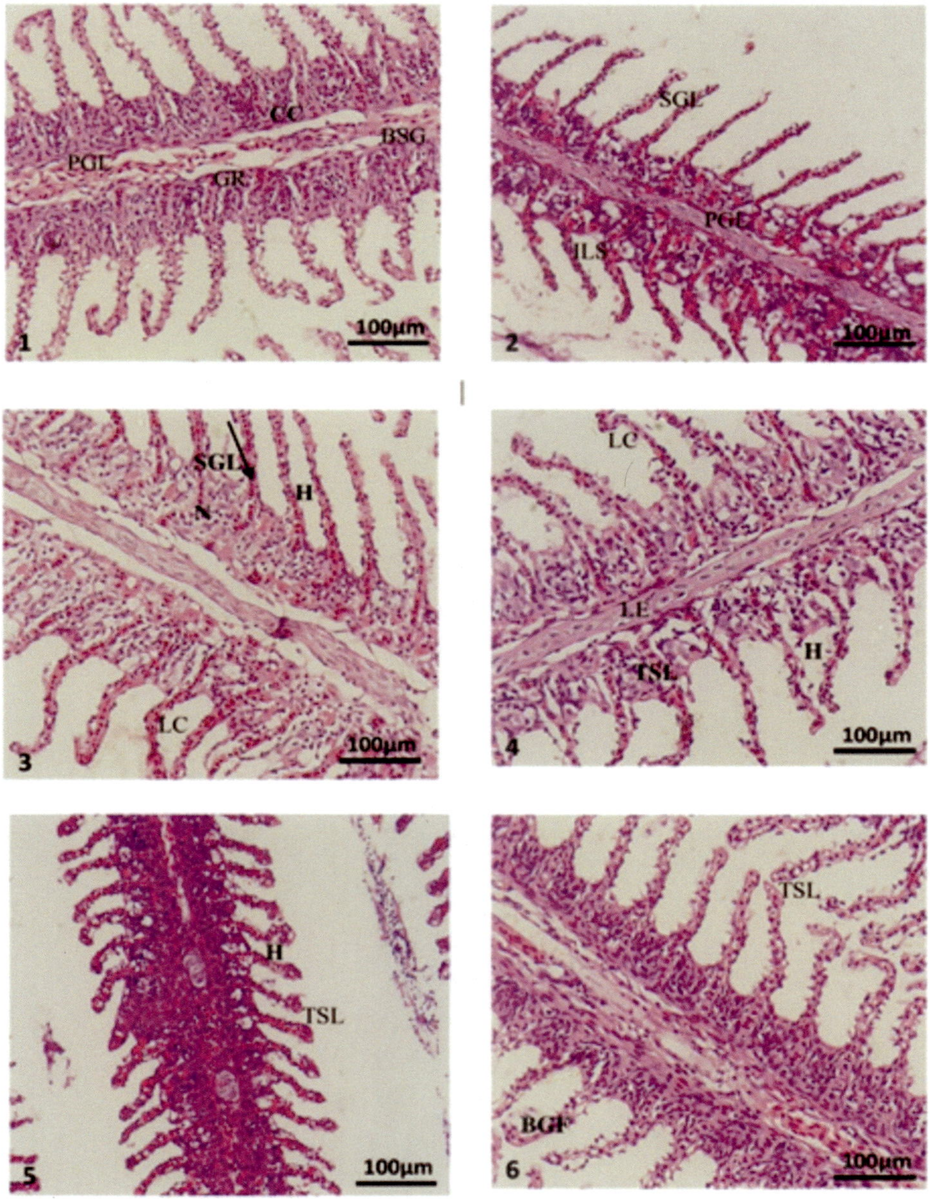
CC
BSG
PGL
GR
100µm
1
SGL
PGL
ILS
100µm
2
SGL
N
H
LC
100µm
3
LC
LE
TSL
H
100µm
4
H
TSL
100µm
5
TSL
BGF
100µm
6

Plate 1 (Pg 6)

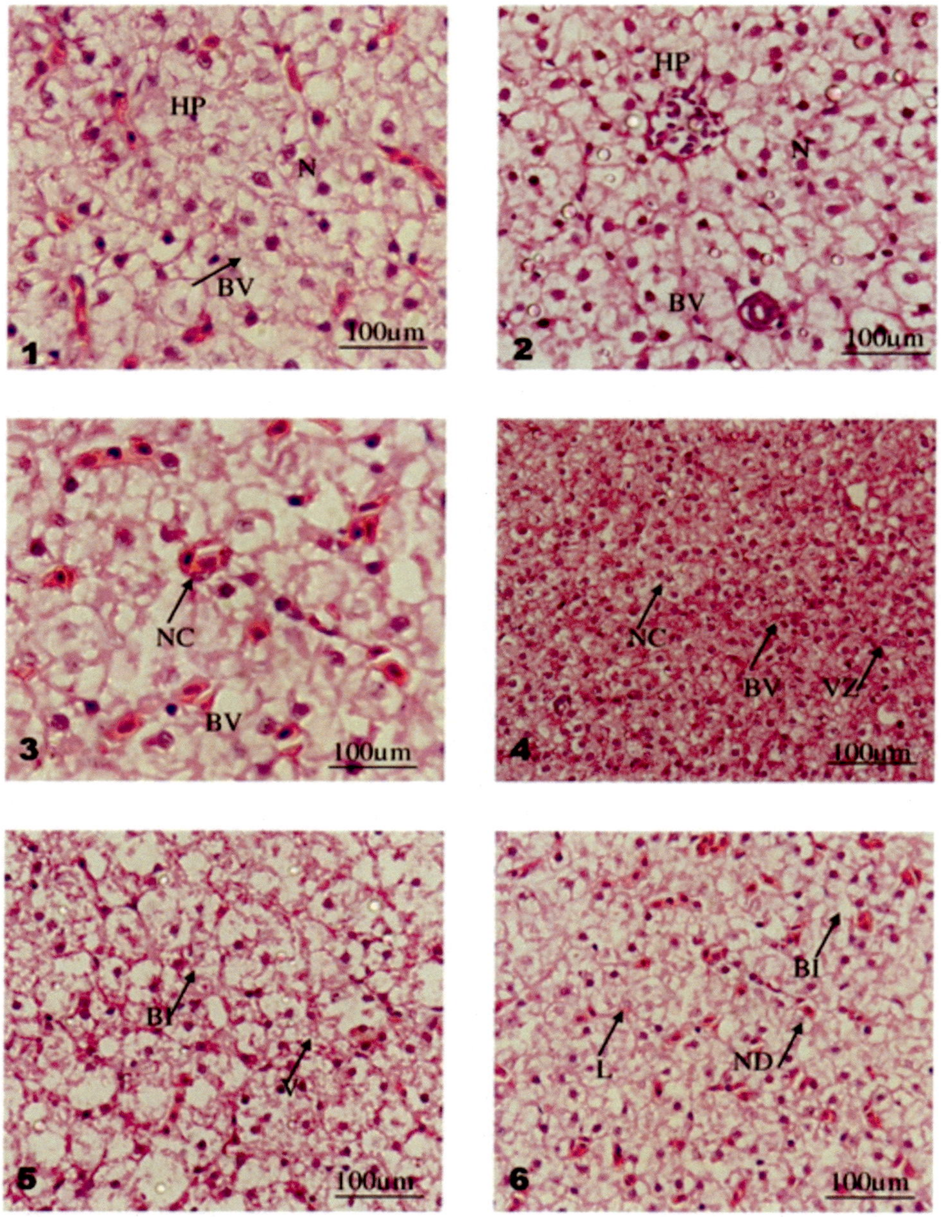
HP
N
BV
100um
1
HP
N
BV
100um
2
NC
BV
100um
3
NC
BV
VZ
100um
4
BI
V
100um
5
BI
L
ND
100um
6

Plate 2 (Pg 8)

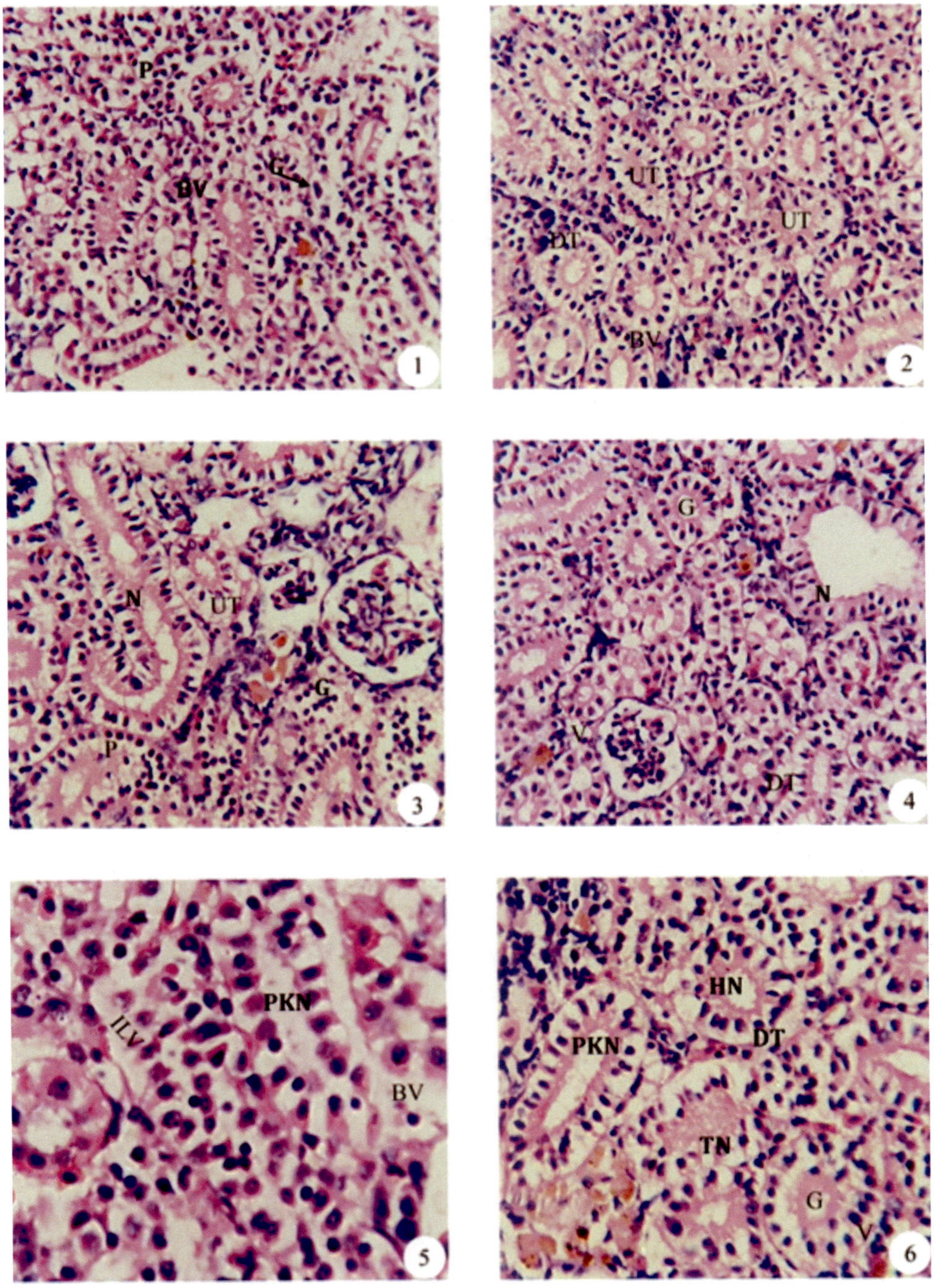

Plate 3 (Pg 10)

Fig 1: Formation of brown colour indicates Zno nanoparticles (Pg 24)

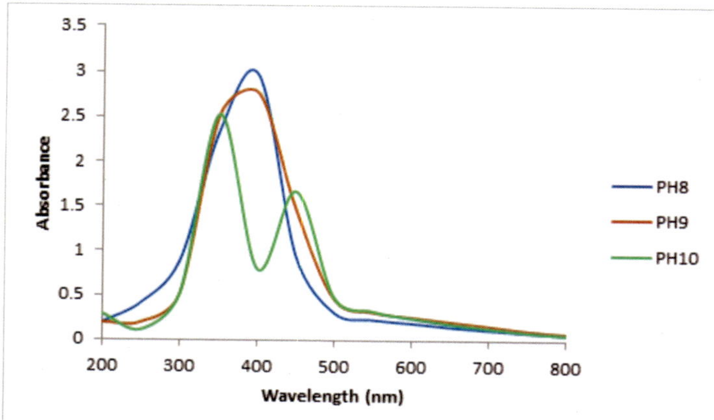

Fig 2: UV-visible analysis of ZnO NPs synthesized from OC plant extract (Pg 24)

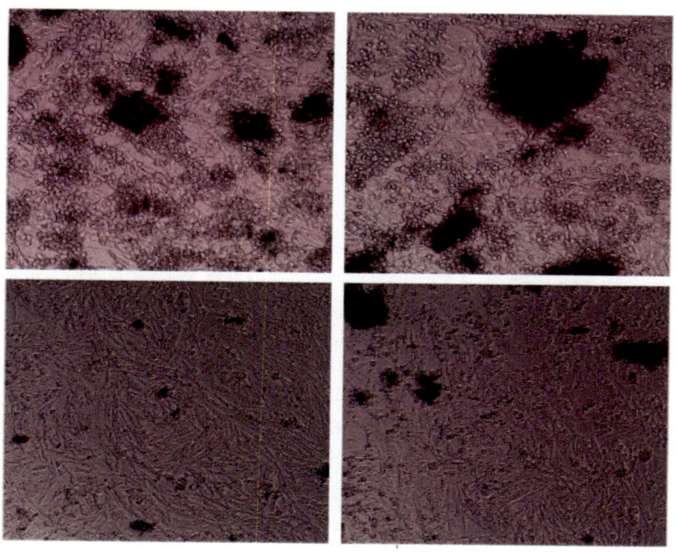

Fig. 8: Images of anticancer activity (Pg 29)

CHAPTER 3

THE ROLE OF ARBUSCULAR MYCORRHIZAL FUNGI WITH PRESSMUD AND GROWTH REGULATORS ON TOMATO

Pushpa K Kavatagi[1], Avitha K. Marihal[2] and H C Lakshman[3]

[1]Vijayanagar College, Hosapete
Email: pushpak.k2@gmail.com
[2]BVVS Akkamahadevi Women's Arts, Science, Commerce College, Bagalkot-587 101
Email: dravikm@gmail.com
[3]Rtd Professor P.G Studies in Botany, Karnatak Science College, Dharwad-580001

Abstract:

Arbuscular mycorrhizal fungi (AMF) establish symbiotic associations with most terrestrial plants. These soil microorganisms enhance the plant's nutrient uptake by extending the root absorbing area. AM fungi also interact with most crop plants including cereals, vegetables, and fruit trees; therefore, they receive increasing attention for their potential use in sustainable agriculture. AM Fungi are commonly known as bio-fertilizers. The present investigation was under taken to understand the effect of AM fungi, Pressmud and Indole acetic acid on biometric yield parameters of Tomato (*Solanum lycopersicum* L., Var NS 585) plants. Pressmud containing 10% organic carbon and it converted into organic manure by the beneficial microbial activity. The plants treated with growth hormones increased the seed yield and remained physiologically more active preserve the sufficient food materials for developing flowers and seeds. Thus, a plant colonized by AM Fungi will use more of these adaptation mechanisms compared to a plant without mycorrhizae. It can be concluded that inoculation of AM fungus alone had significant growth parameters compared to plants grown with amendments of pressmud and IAA alone, but these results are more significant over the control treatment.

Key words: *Arbuscular mycorrhizal fungi, Nutrient uptake,Pressmud, Vegetables and interaction.*

Introduction

Frank was coined the word mycorrhizae he is a German researcher; it is originated from the Greek (mycos- fungus and rhiza-root). Mycorrhiza is a symbiotic relationship between soil fungi and plant roots it is neither the fungus nor the root, but rather the structure formed from these two partners. Since, the association is mutualistic and both organisms benefit from the association. The fungus receives carbohydrates and growth factors from the plant, which in turn receives many benefits, including increased nutrient absorption. In this association, the fungus takes over the role of the plant's root hairs and acts as an extension of the root system. Maize (*Zea mays L.*) is one of the most important food crops of the world. It has a remarkable productive potential along with other members of the family Poaceae, such as wheat and rice (Kling and Edmeades 1997).

Arbuscular mycorrhizal (AM) fungi are native to agricultural soils and form a mutualistic symbiosis with the majority of crop plants. Among the benefits to the plant ascribed to the symbiosis are enhanced: uptake of immobile nutrients, water relations, and disease resistance (Smith and Read, 2008). These benefits make utilization of the symbiosis attractive to sustainable agricultural systems that are designed to minimize synthetic inputs of fertilizer and pesticides, or in the case of organic agriculture, eliminate them from the production system. The role of arbuscular mycorrhiza (AM) in enhancing plant growth and yield of crops has been previously reported (Bolan, 1991). In this regard one of the greatest potential benefits of AM fungi for host plants is the increase in biomass and growth of the host plants. This increase in growth and biomass may be caused by the host plant's increased ability to acquire essential nutrients and water (Ruiz-Lozano & Azcon, 1995).

Tomato (*Solanum lycopersicum* L.) belongs to family Solanaceae. It was originated in western South America. Tomato is an herbaceous perennial but is usually grown as an annual crop in temperate regions. The fruit is a berry containing many seeds, the presence of carotenoid and lycopene tomato varieties have red fruits. Tomato is in a third rank by weight in global production of vegetables and tomato fruits or products are widely consumed all over the world (Tan *et al.*, 2010), tomato is rich source of vitamin C, vitamin A and protein. Tomato juice contains amino acids and glutamic acids, Tomato (*Solanum lycopersicum* L.) is one of the world's major vegetables with 4.4 million ha under production and 115 million tons produced worldwide in 2004 (FAO STAT data, 2004). Press mud is latent quality waste source for hydrogen production. About 8% of crushed sugar cane is converted in to press mud, in India annually about 4.2 million tons of press mud is available from the sugar industry reported by (Saravanane and Radjaram, 1998). About 12 million tons of pressmud, a major byproduct of sugarcane processing, is produced annually (Parthasarathi, 2007). Pressmud is a valuable resource of plant nutrients therefore it affects the physical, chemical and biological properties of a soil (Rangaraj *et al.*, 2007). It consists of residue such as fiber, ash, soil, and other particles in cane juice obtained from sedimentation of suspended materials. In vermicompost the pressmud is used as a suitable substrate have been observed by (Parthasarathi and Ranganathan, 2002). Pressmud from the sugar mills is a very useful source of fertilizer as well as for some chemicals. The developmental regulation of the mycorrhizae may be achieved through modification of endogenous phytohormone levels and through the regulation of phytohormone receptors in the fungus and for their host. For sustainable production the pressmud is one of the sources of organic matter, it is neither a chemical fertilizers nor organic manures, but their integrated use has been observed to favorably beneficial and it can be utilized to enhance the crop production and nutrient status of the soil. The process of adding pressmud to the soil will increase microbial activity in the soil it gives intellectual benefit as it is considered as an effective sources of carrier material for inoculation of microbes.

Arbuscular mycorrhizal colonization of the roots results in formation of intracellular exchange interfaces in root cortical cells, which are composed of the membranes of both symbionts together with an interfacial apoplast between them. Outside the root a network of extraradical mycelium develops, extending beyond the rhizosphere and increasing the volume of soil from which essential nutrients are captured by an AM plant. Mycorrhizal plants can absorb and accumulate phosphate and other micronutrients from the soil or solution several times more than non-mycorrhizal plants. This advantage of mycorrhizal association helps in reducing the chemical fertilizer requirement of phosphorus and some micronutrients such as zinc. Under optimal conditions, mycorrhizal association can replace the requirement of chemical phosphorus up to 25 percent These mycorrhizal associations also ensure better utilization of applied phosphorus, which can be quickly fixed in insoluble form in certain soils. A common effect of drought stress, similarly as of other environmental stresses, is to cause oxidative damage (Smirnoff 1998). Generation of reactive oxygen species (ROS) lead to lipid peroxidation (Chen et al. 2000, Sreenivasulu et al. 1999) protein degradation (Jiang and Zhang 2001). Compared to uninoculated plants, AMF inoculation had positive effect on the growth of carrot and sorghum (Kim et al. 2017). The use of microbial inoculants plays a vital role in sustainable agricultural fungi are known to improve the nutritional status, growth, and development of plants, protect plants against root pathogens and resistance to drought and salinity. AM Fungi have been shown to increase resistance to root infecting pathogenic fungi and bacteria. In recent years the potential role of AM Fungi in helping to increase a plants tolerance and uptake of heavy metals for the restoration of degraded natural ecosystems has been investigated. Though AM fungi are not host specific, recent studies have clearly brought out host preference in AM fungi thus clearly defined the need for selecting efficient AM fungi for inoculating a particular host.

Materials and Methods

Soil and plant material

The characteristic of soil physical and chemical used for pot experiments were estimated according to Jackson, (1973) and it shown in table 1.1. The soil: sand (3:1 v/v) was filled into 18 cm diameter pots and filled with 3 kg of soil, the seeds of PMK 1, Vaibhav, NS 524 and NS 585 varieties of *Solanum lycopersicum* L., were procured from Namdhari seed company Bangalore, India. Seeds were surface sterilized by treating with 1% sodium hypochlorite for 2-3 min before sowing and after germination uniform seedlings were selected for further experiments.

Inoculum

The *Rhizophagus fasciculatus* (Thax.)Walker&Schußler were mass multiplied in 18 cm diameter containing 8 kg sterilized sand : soil (1:1 v/v) mixture as the substrate and *Sorghum vulgare* L., (Jowar) as host plant. After 60 days of growth shoots of Jowar were chopped and the inoculum containing spores, root bits was air dried and 10 g of the mycorrhizal inoculum was applied to the planting area at a depth of about 4 cm to the pots.

Pot experimental setup

The experimental pots amended with air-dried sterilized soil-sand mix (3:1 v/v). Ten grams of AM fungal (*Rhizophagus fasciculatus*) inoculum consisted of colonized roots and soil contained 115-172 chlamydospores, as single inoculation, pressmud was applied to the planting area depth of about 5 cm to the pots before sowing seeds and again it was treated with 2% of IAA (Indole acetic acid) hormone as follows.

A. Control

B. *Rhizophagus fasciculatus* (Thax.)Walker&Schußler

C. Pressmud

D. Indole acetic acid (IAA)

E. *R. fasciculatus (Rf)* + Pressmud+ Indole acetic acid (IAA)

The plants were uprooted periodically at 30, 60 and 90 days. The growth parameters such as root length, fresh weight of root, dry weight of root, fresh weight of shoot, dry weight of shoot, per cent root colonization, spore number and number of leaves, were recorded . The AM fungal spores were counted following the procedure by wet sieving and decanting technique (Gerdemann and Nicolson, 1963). The per cent of mycorrhizal colonization was evaluated microscopically followed by clearing of roots in 10% KOH, neutralized in 2% HCl and stained with 0.05% trypan blue in lactophenol according to the method described by Phillips and Hyman, (1970).

$$\text{Percent of root colonization (\%)} = \frac{\text{No of root bits colonized}}{\text{Total number of root bits observed}} \times 100$$

Statistical Analysis

Data were subjected to analysis of variance and mean values were compared using the Duncan's multiple range test (at the 0.05 level), as recommended (Snedecor, and Cochran, 1980).

Result

The inoculation of arbuscular mycorrhizal fungi and pressmud with Indole acetic acid augmented plant growth (Figure 1), and the growth parameters were recorded in the experimental variety. The results of NS 585 varieties of *Solanum Lycopersicum* L. are described below.

Experimental results showed that after 30 days, plants grown amendment pressmud and IAA with AM fungus inoculation showed significantly increased growth parameters over the other treatments (Table 2). Increased shoot length (12.52 cm), fresh (8.45g) and dry weight of shoot (2.11g), root length (3.50cm), fresh (1.83g) and dry weight of root (0.25g), number of leaves (27.66),per cent root colonization (42.00%), spore number (50.00) 50 g of soil was associated with plants exposed to pressmud and IAA in presence of AM fungus inoculation.

Similar trend was observed in experimental plants after 60 days, of plant growth to the pressmud amendment and inoculated with AM fungus increased shoot length (40.10cm), fresh (69.53g) and dry weight of shoot *Rhizophagus fasciculatus*, in the pressmud and IAA there was a significant increase in (5.56g), root length (8.11cm), fresh (13.83g) and dry weight of root(2.89g), number of leaves (71.66), number of flowers (6.66), number of fruits (3.33), root colonization (63.33%), spore number(67.33), stem diameter (2.08cm).

At 90 days plant grown with amendment of pressmud, and IAA with AM fungus resulted significantly increased shoot length (45.10cm), fresh (125.91g) and dry weight of shoot (12.14g), root length (15.49cm), fresh (24.55g) and dry weight of root (3.23g), number of leaves (102.00), number of flowers (3.66), number of fruits (6.66), root colonization (86.00%), spore number (99.33) in 50 g of soil, stem diameter (3.34cm) compared to plants grown with other treatments. Inoculation of AM fungus alone had significant growth parameters compared to plants grown with amendments of pressmud and IAA alone, but these results are more significant over the control treatment shown in (Figure 2).

Discussion

Based on the experimental results and observation on Tomato varieties with the inoculation of AM fungi would have the advantage of reducing the amount of fertilizers that must be added. The shoot length, fresh and dry weight of shoot and root, number of flowers, fruits and leaves, per cent of root colonization, spore number 50 g-1 of soil and stem diameter of NS 585 varieties of *Solanum lycopersicum* L., were enhanced with the inoculation of AM fungi (*Rhizophagus fasciculatus*) than noninoculated control plants. *Rhizophagus fasciculatus*, pressmud and indole acetic acid increased the plant growth. There was an increased shoot and root dry biomass production in response to mycorrhizal inoculation AM fungus (*Rhizophagus fasciculatus*).

Although a number of case studies on use of pressmud are available our investigation was undertaken to understand the effect of AM fungi, Pressmud and Indole acetic acid on biometric yield parameters of Tomato (*Solanum lycopersicum* L.,) plants. A pot culture experiment was conducted in green house, using three combinations: AM Fungi, Pressmud and indole acetic acid. The plants grown in pots were investigated to evaluate the effects of *Solanum lycopersicon* L., (Var. NS 585). Growth regulators significantly influenced various physiological and biochemical parameters. Plant growth and biomass was significantly increased to all the treatments of AM Fungi, Pressmud and indole acetic acid treatment compared to non-inoculated (control) plants. This was found to be superior in augmenting the growth and yield of tomato plants (Pushpa K. and H.C. Lakshman, 2014). Similar findings were reported by many researchers (Nehra and Hooda, 2002; Naik and Rao, 2004), they have reported that application of pressmud to the lentil crop increased plant height. Inoculation of mycorrhizal fungi provide direct effect between soil and roots and reduce the undesirable effect of heavy metal toxicity (Leyvel *et al.*, 1997). The inoculation of *Rhizophagus fasciculatus* with pressmud was synergetic effect in reduction of nematode diseases and increased the plant growth and yield, it induced systemic resistance in nematode.

Table 1 revels the experimental results on effect of *Rhizophagus fasciculatus* (Thax.) Walker & Schußler, Pressmud and Indole acetic acid on growth characteristics of *Solanum lycopersicum* L. var. NS 585 for 30, 60, and 90 days. References

Table 1: Effect of *Rhizophagus fasciculatus* (Thax.) Walker & Schußler., Pressmud and Indol acetic acid on growth characteristics of *Solanum lycopersicum* L. var. NS 585 for 30, 60, and 90 days.

Treatments	Shoot length (cm)	Fresh weight of shoot (g)	Dry weight of shoot (g)	Root length (cm)	Fresh weight of root (g)	Dry weight of root (g)	Number of leaves	Number of flowers	Number of fruits	Percent of root colonization (%)	Spore number	Stem diameter (cm)
30 Days												
CN	3.50±0.08e	0.35±0.02e	0.09±0.05d	1.51±0.08e	0.29±0.08d	0.07±0.02d	9.00±0.57d	0.00±0.00e	0.00±0.00e	0.00±0.00c	0.00±0.00c	0.49±0.01c
Gf + Press-mud + IAA	12.52±1.45a	8.45±0.01a	2.11±0.05a	3.50±0.08a	1.83±0.01a	0.25±0.01a	27.66±0.33a	0.00±0.00a	0.00±0.00a	42.00±1.52a	50.00±1.52a	1.59±0.02a
Rhizophagus fasciculatus	10.08±0.04b	7.07±0.02b	1.31±0.04b	3.09±0.05b	0.91±0.08b	0.15±0.02b	22.33±0.88b	0.00±0.00b	0.00±0.00b	32.00±0.57b	36.33±0.88b	1.08±0.04b
Pressmud	8.08±0.04c	4.10±0.02c	0.84±0.01bc	2.52±0.01cd	0.55±0.02c	0.11±0.08b	18.33±0.33c	0.00±0.00c	0.00±0.00c	0.00±0.00c	0.00±0.00c	1.04±0.02b
IAA	7.52±0.01d	3.28±0.01d	0.18±0.02c	2.11±0.05d	0.32±0.01d	0.10±0.02cd	16.66±0.33c	0.00±0.00d	0.00±0.00d	0.00±0.00c	0.00±0.00c	0.51±0.08c
60 Days												
CN	18.08±0.04e	25.17±0.01e	1.31±0.05c	4.51±0.08e	6.24±0.02d	0.71±0.08e	36.14±0.35c	1.66±0.33c	1.25±0.3c	0.00±0.00c	0.00±0.00c	1.00±0.15c
Gf + Press-mud + IAA	40.10±0.05a	69.53±0.08a	5.56±0.02a	8.11±0.05a	13.83±0.01a	2.89±0.02a	71.66±1.20a	6.66±0.33a	3.33±0.66a	63.33±1.20a	67.33±1.20	2.08±0.04a
Rhizophagus fasciculatus	32.09±0.04b	46.53±0.01b	2.94±0.02b	7.51±0.08b	10.85±0.01b	1.52±0.01bc	64.00±0.57ab	3.66±0.33b	2.66±0.33ab	45.00±1.73b	42.33±1.45b	1.52±0.01b
Pressmud	27.11±0.05c	41.11±0.02c	2.52±0.01bc	6.12±0.06cd	9.18±0.08c	1.29±0.04c	55.00±0.57b	2.66±0.33bc	1.66±0.33b	0.00±0.00c	0.00±0.00c	1.18±0.04b
IAA	25.09±0.05d	38.47±0.03d	1.66±0.05c	5.37±0.11de	7.81±0.08cd	0.82±0.08de	52.66±0.88b	1.66±0.33c	1.53±0.33b	0.00±0.00c	0.00±0.00c	1.11±0.06b
90 Days												
CN	25.10±0.09e	36.15±0.02d	5.17±0.03de	7.54±0.02e	9.91±0.03e	2.14±0.02d	50.00±0.57d	1.33±0.33d	2.33±0.33d	0.00±0.00c	0.00±0.00c	1.51±0.08d
Gf+Press-mud+IAA	45.10±0.05a	125.91±0.01a	12.14±0.02a	15.49±0.07a	24.55±0.02a	3.23±0.04a	102.00±0.57a	3.66±0.33a	6.66±0.33a	86.00±0.88a	96.33±0.88a	3.34±0.17a
Rhizophagus fasciculatus	43.09±0.04b	55.83±0.01b	7.54±0.02b	11.52±0.01b	15.14±0.02b	3.23±0.05ab	71.33±0.88b	2.66±0.33ab	3.66±0.33b	68.00±0.57b	81.33±0.53b	2.50±0.01b
Pressmud	37.12±0.06c	51.64±0.02c	4.42±0.01c	9.50±0.05c	13.61±0.03c	2.71±0.08c	65.00±1.73c	1.88±0.33bc	2.66±0.33cd	0.00±0.00c	0.00±0.00c	2.54±0.02b
IAA	32.12±0.06d	45.23±0.03c	4.25±0.07d	8.12±0.06d	11.19±0.08d	2.46±0.02cd	63.00±0.57c	1.66±0.33bc	1.66±0.33d	0.00±0.00c	0.00±0.00c	2.02±0.02c

CN - Control, IAA- Indol acetic acid all the value represents. Data represents means ± SE of 3 replicates. Mean separation within column by Duncan's multiple range test at $P < 0.05$.

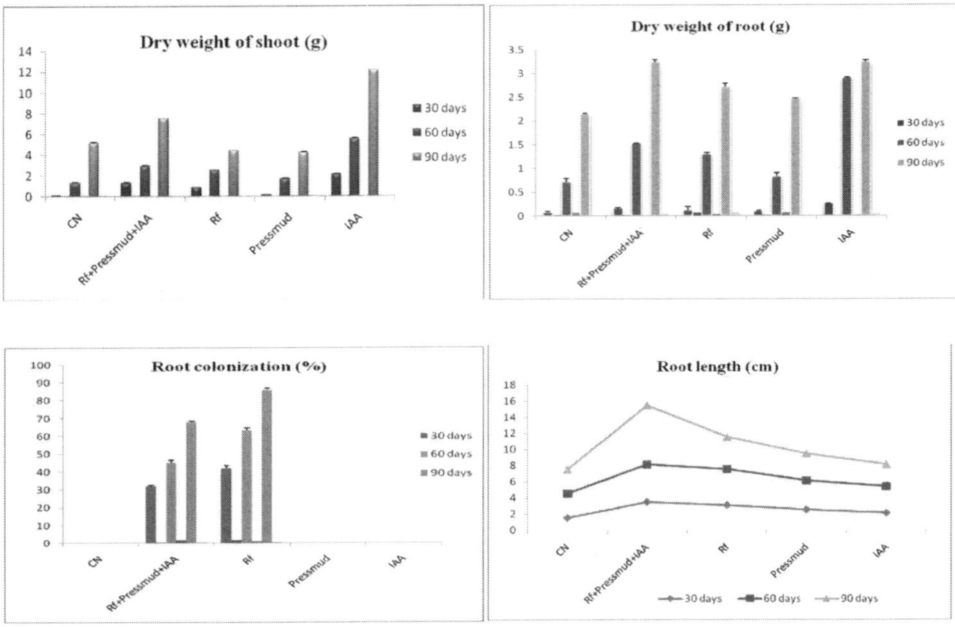

Fig 1. Effect of *Rhizophagus fasciculatus*, Pressmud and IAA on dry weight of shoot, dry weight of root, per cent root colonization and root length in *Solanum lycopersicum* L., (Var. NS 585) at 30, 60 and 90 days.

Conclusion

It was reported positive efficiency of mycorrhizae to protect plants from abiotic and biotic stress conditions and to supply to plant nutrition's with a lot of studies. Arbuscular mycorrhizas have the ability to alleviate many anthropogenic stresses, including effects of metals and polychlorinated aliphatic and phenolic pollutants. Remediation of soils contaminated with toxic pollutants using specific combinations of plants and species or strains of mycorrhizal fungi deserves increased emphasis.

The findings of the present study reference to the organic substrate, pressmud – a sugar industry waste by-product. A combination of pressmud and IAA with inoculation of Rhizophagus fasciculatus, one could positively derive wealth from the sugarcane and sugar industry wastes. The application of fresh pressmud vermicasts with increased microbial population and activity to the soil support better improvement of soil fertility. The findings of the present study suggest that qualitative differences rather than nutrient intensity determine the potential to rely on AM fungi in plant P nutrition in contrasting cropping systems. The present study clearly indicates that Solanum lycopersicum L., is a rich source of phytoconstituents having immense antioxidant potential.

Acknowledgement

This paper and the research would not have been possible without the continual support of our family. I would like to thank our colleagues and friends for their insightful views about our research work. Author is grateful to the UGC, New Delhi for providing financial assistance. Our heartfelt gratitude goes to our research guide Dr.H.C.Lakshman for his persistent guidance, encouragement and support.

References

N.S. Bolan,A critical review on the role of mycorrhizal fungi in the uptake of phosphorous by plants. Plant Soil 134, 189-207, (1991).

S.E Dale, A .Doherty-Kirby, G.Lajoie and D.E .Heinrichs. Role of siderophore biosynthesis in virulence of Staphylococcus aureus: identification and characterization of genes involved in production of a siderophore. Infect. Immun. (2004). 72, 29–37.

FAO STAT data. Accessedat: collections version=ext & has bulk=0&subset=agriculture(2004).

E.Giovannucci., A .Ascherio., E.Rimm ., Stampfer, M., Colditz, G., Willett, W. Intake of carotenoids and retinol in relationship to risk of prostate cancer. J Natl Cancer Inst (1995). 87:1767–1776.

J.G .Kling and G. Edmeades., Morphology and growth of maize IITA/CIMMYT. Research guide 9 Training programme. International institute of tropical agriculture IITA, Ibadan, Nigeria, (1997).Pp31.

K.Parthasarathi., Influence of moisture on the activity of Perionyx excavates (Perrier) and microbial–nutrient dynamics of pressmud vermicompost. Iranian J. Environ. Health Sci. Engg. (2007) 4 (3), 147–156.

K.Parthasarathi., L.S Ranganathan., Supplementation of pressmud vermicasts with NPK enhances growth and yield in leguminous crops (*Vigna mungo* and *Arachis hypogaea*). J. Curr. Sci. (2002). 2, 35–41.

Pushpa K. Kavatagi and H.C. Lakshman., Effect of Arbuscular Mycorrhizal Fungi, Pressmud and Growth Regulator on *Solanum lycopersicum* L. Int. J. Pure Appl. Sci. Technol., (2014). 21(2), pp. 11-16.

T Rangaraj., E.M. Somasundaram, M. Amanullah, V. Thirumurugan, S. Ramesh and S. Ravi.,. Effect of agro industrial wastes on soil properties and yield of irrigated finger millet (*Eleusine coracana* L. Gaertn) in coastal soil. Res. J. Agric. & Biol. Sci. (2007).,3(3): 153-156.

J.M .Ruiz-Lozano., & R .Azcon.,. Hyphal contribution to water uptake in mycorrhizal plants as affected by the fungal species and water status. Physiologia Plantamm. (1995) 95, 472-478.

R .Saravanane., B Radjaram., Integrated biomethanation system for waste management of sugar industry. J. Inst. Public Health Eng. (1998)., (1), 22–29.

S. E .Smith., and D. J. Read.,. Mycorrhizal symbiosis, 3rd ed. Academic Press, London, UK. (2008).

CHAPTER 4

ACUTE ORAL TOXICITY STUDY OF "POLYHERBAL FORMULATION (*ROSMARINUS OFFICINALS*, ASHWAGHANDHA & AMLA) IN SPRAGUE DAWLEY RATS"

[*1]Dr. Shivaraj Y and [2]Dr. Rajesh R

*[*1]Assistant Professor, Department of Zoology, L.V.D College Raichur-584 103*
[2]Chromed Biosciences Pvt. Ltd. A Multi-Division Contract Research Organization, KIADB Industrial Area, Hirehalli, Tumakuru-572 168, Karnataka, India.
Email: shivarajmetigouda@gmail.com, Mobile: +91-8971316472

Abstract:

Today in this present globe many part of the continents the use of herbal medicinal products and supplements has increased tremendously over the past three decades with not less than 80% of people worldwide trusting on them for some part of primary healthcare. The study was focused to evaluate the acute oral toxicity of polyherbal formulations (PHFs) and the developed polyherbal formulation includes three herbs viz. the roots of Withania *somnifera* (Ashwagandha), *leaves of* R. *officinalis* (Rosemary) and fruit of Emblica *officinalis* (Amla) detect the effect of this polyherbal formulation on higher mammals. The raw materials were standardized according to the WHO guidelines and the three holistic extracts were mixed in 1:1:1 ratio for further study by bio-guided method and also in combination to control and curing or managing of different diseases.

However, to confirm the drug's safety and efficacy in the appropriate dosages we used healthy nulliparous and non-pregnant female Sprague Dawley rats (Rattus *norvegicus*) with average body weight of 160 g - 180 g and age between 8–12 weeks were used throught the experiments in the present study administered a single dose of 2000mg/kg of body weight by oral gavage in female Sprague Dawley rats the observation period of 14 days. The rats were maintained under standard laboratory conditions (room temperature of 22.1 to 24.6 C with optimal air cycle changes 12-15 per hour and relative humidity of 48-61 % with 12 hours light & dark cycle). The animals were fed with laboratory animal feed and UV sterilized drinking water was provided *ad-libitum* throughout the experimental period. Hence the present study is designed to test the acute (limit dose) toxicity of two herbal extracts of rosemary leaf and ashwagandha and amla root in Sprague Dawley rats following the OECD guidelines 423, which would help in suggesting the minimum safety level of herbal extracts.

Keywords: Acute oral toxicity, Polyherbal formulation, Bio-guided method, *ad-libitum* & Sprague Dawley rats.

INTRODUCTION

Today in this green planet the Plants are very useful to mankind. Many of them are used exclusively for medicinal purposes. According to the World Health Organization (WHO), "a medicinal plant is a plant which, in one or more of its organs, contains substances that can be used for therapeutic purposes, or which are pioneers for chemo-pharmaceutical semi-synthesis." Such plants are in great demand by pharmaceutical companies for their active ingredients. Herbal care or traditional systems of medicine are used from ancient times; herbs have been the original source for most of the drugs. Today approximately 70% of the world population is depending on medicinal herbs. Medicinal plants contain so many chemical compounds which are the major source of therapeutic agents to cure human diseases and Herbal formulations have attained wide recognition in comparison to crude plant materials and extracts, due to reduction in dose levels, convenience and ease of administration. These formulations are popular worldwide as therapeutic agents, in various ailments that impact the quality of life of human beings. Recent research focus has shifted towards the use of herbal medicines due to their diverse biological activities, easy availability, cost effective in nature and safe usage. Polyphenols and flavonoids present in the medicinal plants have been shown to possess significant anticancer properties. The information of extractability of a plant provides an idea regarding the amount of extract present in a definite quantity of drug. The extractability also serves as a tool for quality control of plant-drug by [1] today the Herbal medicines are traditionally given in the form of polyherbal formulations (PHFs) as each ingredient is supposed to have different pharmacological function. Polyherbal formulations are usually prescribed to be taken for a longer period and hence may cause adverse effects in the patients, thereby deserving evaluation of their efficacy and safety profile.

A World Health Organization survey indicated that 70 to 80% of the global population depends on alternative medicine, predominantly herbal in nature, in their primary health care. The uses of medicinal plants as a source of drugs in primary health care have become popular universally, particularly in developing countries as a safe because of natural source. [2] Despite the widespread use of plants for treatment of several ailments there is a little known about their toxicity and safety. However, in contrast to popular view, there are some reports regarding serious adverse effects of herbal therapies, such as renal failure and liver injury caused by some plant species. In this sense, experimental studies to determine the safety of medicinal plants are required. The latest surveys have indicated that many medicinal plants applied in traditional medicine showed adverse effects. Therefore, it should be emphasized that the traditional use of any plant for medicinal purposes, by no means, guarantees the safety of such plant. Therefore, the evaluation of the toxic action of the plant extracts or herbal formulations is important in order to consider them safe before used as medicines.

A key stage in ensuring the safety of drugs is to conduct toxicity tests in appropriate animal models. Toxicity is the fundamental science of poisons. The Organization for Economic Co-operation and development (OECD) mentioned acute toxicity as the advance effect occurring within a short time of oral administration of a simple dose of a substance or a multiple doses given within 24 hours and Acute oral toxicity data is a key component of classification and labeling as well as hazard assessment for industrial substances and is a requirement in diverse regulatory testing regimes. The first acute oral toxicity test was created by the British pharmacologist by [3]. It used to involve up to 150 animals until the OECD Test Guideline (TG) 401 [4] standardization it to 45 animals in the year 1981 by [5]. The acute oral toxicity test aims at establishing the therapeutic index, i.e. defined as the ratio LD_{50}: ED_{50}. However, the term acute oral toxicity is most often used in connection to lethality and lethal dose determinations. It is used to standardize biological products and can serve to establish dosing levels for repeated dose studies. With this keeping in view the impact of stress and the great demand for effective supplement to overcome stress and its consequence, the study was focused to evaluate the acute oral toxicity of polyherbal formulation (PHF).

The developed polyherbal formulation includes three herbs viz. the roots of Withania *somnifera* (Ashwagandha), *leaves of R. officinalis* (Rosemary) and fruit of Emblica *officinalis* (Amla) detect the effect of this polyherbal formulation on higher mammals. Withania *somnifera* (W. *sominifera*), commonly known as Ashwagandha and it is also called as Indian ginseng and winter cherry, holds a key place in Ayurvedic herbs being one of the best researched and effective herbs belonging to family Solanaceae and It is an important medicinal plant in

Ayurvedic and Unani system of medicine. It is widely used indigenous medicine for more than 3,000 years with high medicinal values. Numerous studies indicated immunomodulatory, antidepressive, antioxidant, antitumor, antistress, antiaging, anxiolytic, cardioprotective and neuroprotective activity of W. *Somnifera*. The plant acts as an anti-inflammatory agent through inhibition of complementary system and lymphocyte proliferation by [6]. The European Food Safety Authority (EFSA) has categorized some of the alkaloids of ashwagandha as toxic and harmful. These include anaferine, anahygrine, withanine, sominiferine, somnine, tropine etc. Apart from alkaloids, withaferine A has also been categorized as cytotoxic lactone by EFSA.

Rosmarinus *officinalis* is commonly called Rosemary and it is an evergreen plant of the Lamiaceae (*Labiatae*) family, in folk medicine, rosemary has been used as a choleretic and a diuretic agent. It has been documented to possess a number of therapeutic applications in medicine for curing or managing of different diseases such as DM, respiratory, gastrointestinal disorders, and inflammatory disease. The active constituents of this plant are phenolic diterpenes and triterpenes. The main active compounds of rosemary include carnosic acid (CA), rosmarinic acid (RA), ursolic acid (UA), caffeic acid and carnosol by [7]. Approximately 90% of the total antioxidant activity of rosemary is attributed to carnosol and CA by [8]. *Emblica officinalis* Geart. (E. officinalis) is commonly known as Amla It is also named as *Phyllanthus Emblica or* Indian gooseberry belonging to family Euphorbaceae, is one of the widely used medicinal herb in Ayurveda. *Emblica officinalis* (EO) enjoys a hallowed position in Ayurveda-an Indian indigenous system of medicine. According to believe in ancient Indian mythology, it is the first tree to be created in the universe. It has application as antioxidant, immunomodulatory, antipyretic, analgesic, cytoprotective, antitussive and gastroprotective. Additionally, it is useful in memory enhancing, ophthalmic disorders and lowering cholesterol level, it is useful in memory enhancing, ophthalmic disorders and lowering cholesterol level.

As per the guidelines from Government when any new product is to be put in the market, it is necessary to generate the data regarding its safety. The present investigation is planned with the objective to evaluate the acute oral toxicity of PHF (W. somnifera, *R. officinalis* and E. officinalis) in female Sprague dawley rats and the animals were observed for clinical signs , tremors, Body weight, Behaviour (depression, excitement) and mortality. No mortality was observed in any of the rats. Body weight was recorded at 0,7th and 14th day and all the animals are terminally sacrificed for gross necropsy findings. [9] The method also renders information on the hazardous properties and allows the substance to be ranked and classified according to the Globally Harmonised System (GHS). Detailed toxicological evaluation has to be performed using suitable experimental animals to provide guidelines for selecting a 'safe' dose for human use. In spite of the wide use of "Polyherbal formulation (Rosmarinus *officinals,* Ashwaghandha and Amla) in Sprague dawley rats" in traditional medicine, reports of systematic toxicity studies are not available. Hence, the present study is designed to test the acute (limit dose) toxicity of two herbal extracts of rosemary leaf and Ashwagandha root in Sprague dawley rats following the OECD guidelines, which would help in suggesting the minimum safety level of herbal extracts. To our best of knowledge, no reports are there on the acute toxicity of polyherbal formulation in Sprague dawley rat's studies.

Materials and Methods

A. Distribution & Chemical compositions of all three Species

Acute oral toxicity study of "Polyherbal formulation (Rosmarinus *officinals,* Ashwaghandha and Amla) in Sprague dawley rats" was performed following OECD Guideline 423 in the Chromed Biosciences Private Limited, Hirehalli Tumakuru, India. All the 3 plants belonging to the herbal formulation of R. *officinalis* (Rosemary leaves), root powder of W. *Somnifera* (Ashwaghandha root) and fruit powder of P.*emblica* (Amla fruit) were collected from the local market, Mandipet, Tumakuru, Karnataka, India. The seeds were dried under shade and powdered before use. Plants used in developing the polyherbal formulations for present study are as follows, Rosmarinus *officinalis,* Withania *somnifera* & Embalica *officinalis.* Rosemary is the native of Mediterranean region and is cultivated in Europe and California in US. It is also grown in Algeria, China, Middle East, Morocco, Russia, Romania, Serbia, Tunisia, Turkey, and to a limited extent in India. Temperate climate is suitable for the cultivation of Rosemary. **From fig-1,** Rosmarinus *officinalis* L. is a medicinal plant that belongs to the *Lamiaceae* family and is commonly known as rosemary by [10]. Rosemary is a dense bush, branched, evergreen and blue–white flower, reaching a height of about 1 m by [11]. It is characterized by leaves with 1–4 cm long and 2–4 mm wide, sessile, leathery, linear to linear-lanceolate, with curved edges, dark green upper side and granulosa and page bottom tomentous, with prominent midrib, and very characteristic smell by [12, 13]. Rosemary composed of dried leaves and flowers constitutes a particularly interesting source of biologically active phytochemicals as it contains a variety of phenolic compounds including carnosol, carnosic acid, rosmanol, 7-methyl-epirosmanol, isorosmanol, rosmadial and caffeic acid, with substantial *in vitro* antioxidant activity by [14]. Rosemary has been widely used not only in cooking, especially to modify and enhance flavours, but also in traditional medicine, being a highly appreciated medicinal plant to prevent and cure colds, rheumatism, pain of muscles and joints. **From fig-2,** it is nowadays one of the most popular sources of natural bioactive compounds, and in fact, this plant exerts various pharmacological activities such as antibacterial, antidiabetic, anti-inflammatory, antitumor and antioxidant, among others.

Fig: 1 Rosmarinus *officinalis*

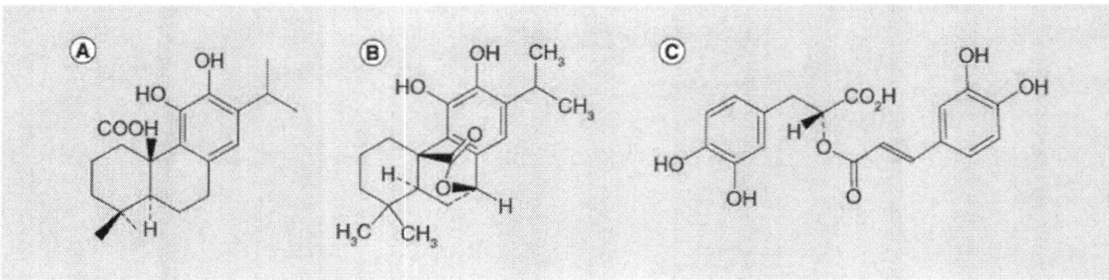

Figure 2. Chemical structure of three major compounds present in *R. officinalis*. (A) Carnosic acid. (B) Carnosol. (C) Rosmarinic acid.

Fig: 2 Chemical Structure

Withania somnifera is a small, woody shrub in the Solanaceae family that grows about two feet in height. It can be found growing in Africa, the Mediterranean, and India. An erect, evergreen, tomentose shrub, 30-150 cm high, found throughout the drier parts of India in waste places and on bunds. Roots are stout fleshy, whitish brown; leaves simple ovate, glabrous, those in the floral region smaller and opposite; flowers inconspicuous, greenish or lubrid-yellow, in axillary, umbellate cymes; berries small, globose, orange-red when mature, enclosed in the persistent calyx; seeds yellow, reniform. The roots are the main portions of the plant used therapeutically. The bright red fruit is harvested in the late fall and seeds are dried for planting in the following spring. It grows in dry parts in sub-tropical regions. Rajasthan, Punjab, Haryana, Uttar Pradesh, Gujarat, Maharashtra and Madhya Pradesh are the major Ashwagandha producing states of the country. From fig-3, Ashwagandha (Withania somnifera, fam. Solanaceae) is commonly known as "Indian Winter cherry" or "Indian Ginseng". It is one of the most important herbs of Ayurveda (the traditional system of medicine in India) used for millennia as a Rasayana for its wide-ranging health benefits. Among the ayurvedic Rasayana herbs, Ashwagandha holds the most prominent place. It is known as "Sattvic Kapha Rasayana" Herb by [15]. Most of the Rasayana herbs are adaptogen / anti-stress agents. Ashwagandha is commonly available as a churna, a fine sieved powder that can be mixed with water, ghee (clarifiedbutter) or honey. It enhances the function of the brain and nervous system and improves the memory. Ashwagandha improves the body's defense against disease by improving the cell-mediated immunity. It also possesses potent antioxidant properties that help protect against cellular damage caused by free radicals. Ashwagandha is one of the most valuable herbal drugs used in Indian traditional medicine

Fig: 3 Withania *somnifera*

(Ayurveda) as a rasayana drug that is capable of imparting long life, youthful vigor, and good intellectual powers. [16] Ashwagandha is clinically used for the treatment of general debility, consumption, nervous exhaustion, insomnia, loss of memory, and so on. [17, 18]. These traditional uses imply that Ashwagandha may possibly be useful at improving neurodegenerative diseases. From fig-4, indeed, this herbal drug has been reported to exert various pharmacological effects such as antiinflammatory, anti-tumor, anti-oxidant, immunomodulatory, and anti-neuropsychiatric disease effects [19, 20]. The biologically active chemical constituents of Withania somnifera (WS) include alkaloids (isopelletierine, anaferine, cuseohygrine, anahygrine, etc.), steroidal lactones (withanolides, withaferins) and saponins [20]. Sitoindosides and acylsterylglucosides in Ashwagandha are anti-stress agents. Active principles of Ashwagandha, for instance the sitoindosides VII–X and Withaferin-A, have been shown to have significant anti-stress activity against acute models of experimental stress [21]. Many of its constituents support immunomodulatory actions [22]. The aerial parts of Withania somnifera yielded 5-dehydroxy withanolide-R and withasomniferin-A [23].

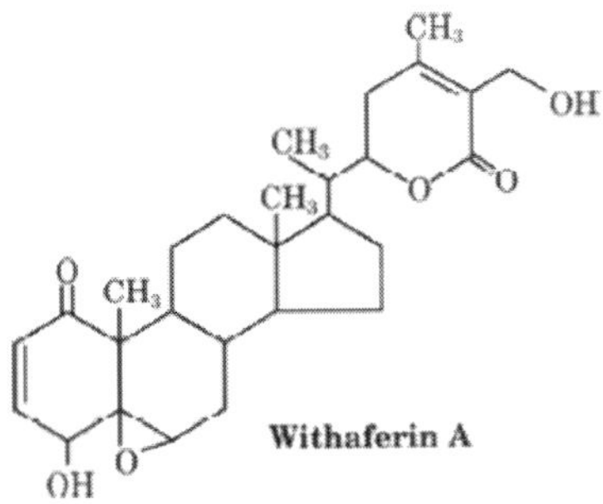

Fig.7: Structure of withaferin A

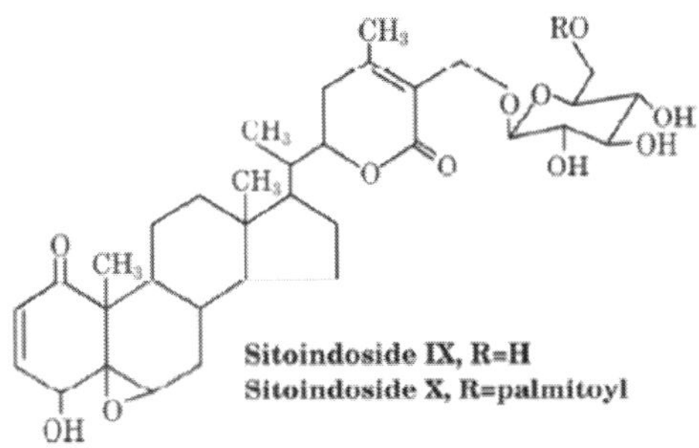

Fig.8: Structure of sitoindosides IX and X

Fig: 4 Chemical Structure

Emblica *officinalis* (also known as Amla or Indian Gooseberry) has an important position in Ayurveda- an Indian indigenous system of medicine. According to believe in ancient Indian Mythology, it is the first tree to be created in universe. It also present on the hill slopes up to 2000 meters. It is found in Pakistan, Uzbekistan, Sri Lanka, South East Asia, China, and Malaysia and commercially cultivated in the state of Uttar Pradesh in India. It is also grown in Tamil Nadu, Rajasthan and Madhya Pradesh by [24]. *Emblica officinalis Gaertn*. (Phyllanthus emblica Linn. Amla. Indian Gooseberry) belongs to the plant family Euphorbiaceae [1]. Amla is a small to medium sized deciduous tree. It grows about 8-18m height with thin light grey bark, leaves are simple, light green, sub-sessile, closely set along the branchlets looks like pinnate leaves; flowers are greenish yellow; fruits are globose, fleshy, pale yellow with six obscure vertical furrows enclosing six trigonous seeds in two seeded three crustaceous cocci [25]. The average weight of the fruit is 60-70 g. The fruits of Amla are widely used in the Ayurveda and are believed to increase defense against diseases. Dried fruits, fresh fruit, seed, leaves, root bark and flowers of Amla are mostly used in medicines [24]. **From fig-5,** *Emblica Officinalis* (Amla), a natural, traditional and functional food in Asia, has physiological benefits such as hepato-, cyto- and radio- protection, as well as hypolipidemic effects. In addition, Amla often functions as a potent antioxidant due to the high level of ascorbic acid (ranging from 1,100 to 1,700 mg/100 g of fruit) in its fruit [26].

The pharmacological research reports on amla reveals its analgesic, anti-tussive, anti-atherogenic, adaptogenic; cardio, gastro, nephro, neuro protective and anticancer properties. From fig-6, Amla is also reported to possess chemopreventive, radio, chemo and immunomodulatory, free radical scavenging, antioxidant, anti-inflammatory, anti-mutagenic activities. These properties are efficacious in the prevention and treatment of various diseases like cancer, atherosclerosis, diabetes, peptic ulcer, anemia, liver, heart diseases and various other disorders [25]. *Emblica officinalis* is one of the most extensively studied plants,primarily it contains tannins, alkaloids, phenolic compounds, amino acids, carbohydrates and other compounds (like Vitamin C, Flavanoids, Ellagic acid, Chebulinic acid, Quercetin, Chebulagic acid, Emblicanin-A, Gallic acid, Emblicanin-B, Punigluconin, Pedunculagin, Citric acid, Ellagotannin, Trigallayl glucose, Pectin). Its fruit juice contains the highest concentration of vitamin C (478.56 mg/100 mL). . Fruits have 28% of the total tannins distributed in the whole plant. The fruit contains two hydrolysable tannins Emblicanin A and B, which have antioxidant properties; one on hydrolysis gives gallic acid, ellagic acid and glucose wherein the other gives ellagic acid and glucose respectively. The fruit also contains Phyllemblin.

Fig: 5 Emblica *officinalis*

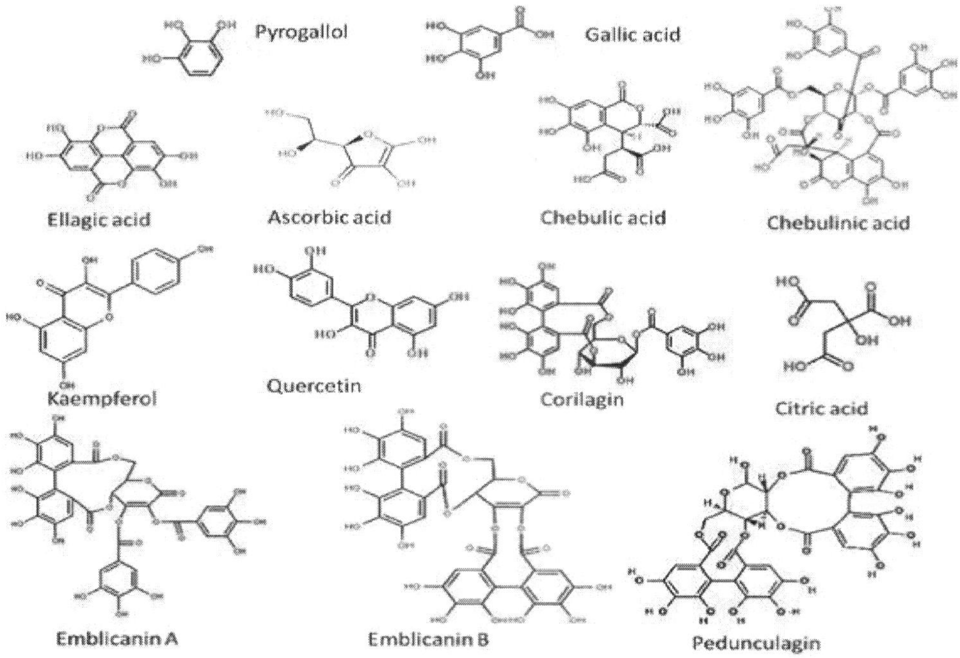

Fig: 6 Chemical Structure

B. Extraction and Preparation of Polyherbal formulation

Extraction of Rosemary leaves: The plant leaves were dried under shade, coarsely powdered and stored in air tight container. This shade dried leaves powder of Rosemarinus officinalis was first defatted with petroleum ether (at 60-80 C) and then extracted 95% v/v ethanol in a sox let apparatus. The alcohol extract filtered, the solvent was evaporated and accurate weight of the extract was taken. The color and constituency of the extract were noted down (light brown and powderly mass), and Extraction of Ashwaghandha root and Amla: Ready dried powder of extracted ashwaghandha root and amla fruit powder were used, and finally Preparation of Polyherbal formulation: The Polyherbal formulation was prepared by mixing all the 3 holistic extracts taking equal Quantity (666mg) of each powdered extract of Rosemary leaves, Ashwagandha root powder and Amla fruit powder were mixed in 10 ml of distilled water and formulated into 1:1:1 ratio for further study.

C. Experimental Animal and their maintenance

A **laboratory rat** or **lab rat** is a rat of the species *Rattus norvegicus domestica* which is bred and kept for scientific research. While less commonly used for research than mice, rats have served as an important animal model for research in psychology and biomedical science. [27].

The rat is the commonly used species for toxicity studies and recommend by the international guidelines (i.e. OECD 423) and it meets the regulatory requirements for conducting preclinical toxicological studies among rodents. Healthy nulliparous and non-pregnant female Sprague dawley rats (Rattus norvegicus) **from fig-7,** with average weight of 160-180 g and age between 8–12 weeks were used for all the experiments in the present study. Before placing animals, the experimental room was decontaminated by fumigation and microbial load was checked by settle plate's method. The copies of results were kept in the raw data. The experimental room floor was mopped with disinfectant solutions daily once. The maximum of 3 animals were housed in a standard polycarbonate cages Size: L43cm X B27cm X H15cm) with stainless steel mesh top grill having facilities for holding pelleted feed and drinking water. The water was provided in polycarbonate water bottles fitted with rubber cork and a stainless-steel sipping tube. Clean sterilized corn cob was provided as bedding material. The corn cob was analyzed periodically for any fungal and microbial contaminations. The animals were fed *ad libitum* with laboratory animal feed and UV sterilized drinking water was provided *ad libitum* throughout the experimental period. The rats were maintained under standard laboratory conditions (room temperature of 22. 1 to 24.6 C with optimal air changes 12-15 per hour and relative humidity of 48-61 %, with 12 hours light & dark cycle).

Fig: 7 Sprague dawley (female)

Animal care and the experimental protocol followed the principles and guidelines suggested by the "Association for Assessment and Accreditation of Laboratory Animal Care" (AAALAC-2012) [28] International accredited facility and registered with Committee for the Purpose of Control and Supervision of Experiments on Animals (CPCSEA-2004) [29] guidelines, Ministry of Environment, Forests and Climate change, Government of India. Also Chromed Biosciences Private Limited ensures that animal experiments are performed in accordance with the recommendation of the guidelines for laboratory animal facility published in the gazette of India, 2018. The care and use of all experimental animals complied with relevant animal welfare laws. Before the experiment animals were allowed to acclimatize for Step I five days and Step II eight days before the experiment (each

Fig: 8 Acclimatization of rat

step 3 rats were used). From fig-8, during acclimatization period, each animal was identified by tail marking with permanent marker pen and cage cards and Clinical sign of individual animals were observed daily during the acclimatization period.

D. Experimental design and Procedures

i. Acute oral toxicity study

This study was performed in accordance with the OECD guideline for the testing of chemicals, "acute oral toxicity study (acute toxic class method)", guideline no. 423, adopted on december 17, 2001 [30]. From fig-9, it is a single dose study for a period of 14 days. Three female Sprague dawley rats were used for each step. The study was conducted in 2 steps. in step-i three female Sprague dawley rats were kept fasting for 16 hrs period of time by providing only water was given ad libitum prior to dosing and the dose volume calculated based on the fasted body weight of each animal and administrated in a single dose of 2000mg/kg of body weight by gavage using a 18g intubation canula. The animal administrated the maximum dose volume were administrated 10 ml/kg body weight. Test item was administrated in constant volume over the ranges of doses to be tested by varying the concentration of dose preparation. The 2000 mg of the test item polyherbal formulation was taken into the mortar and triturate using pestle. 3 ml of distilled water was slowly added in to mortar and mixed. Transfer the formulation to the measuring cylinder and make the volume up to 10 ml. the dosing solutions was freshly prepared prior to dose administration. after test item administration feed was provided at 3rd hour observation and animal were observed periodically for clinical sign of toxicity and body weight for every 30 min of the administration for first 24hrs and then daily for 14 days. No mortality was observed in all the treated animals. Hence step-ii was conducted with three naïve female Sprague dawley rats were treated with same dose of 2000 mg/kg body weight, no mortality was observed in throughout the observation period. Hence further dosing was not performed as per dosing criteria. During treatment period, each animal was identified by body marking with turmeric solution and cage cards and terminally all the animal were subjected to grass necroscopy findings.

ii. Clinical observation and body weight

For acute toxicity study all the animals were observed twice daily for mortality and morbidity throughout the experimental period. Clinical examination included any abnormal physical and behavioural changes and Animals were observed individually for toxic signs and mortality after dosing at least once during the first 30 min, 1, 2 and 4 (± 10 minutes at each point) hours and daily thereafter for a total of 15 days. Observations were include changes in skin and fur, eyes and mucous membranes, and also respiratory, circulatory, autonomic and central nervous systems, and somatomotor activity and behaviour pattern tremors, convulsions, salivation, diarrhoea, lethargy and others and Bodyweight of all animals was measured on day 0,8 and 15. All the animals were normal in behavior throughout the study.

iii. Pathology

At the completion of 14 days of observation period, all the animals were subjected to gross necropsy examination. Microscopic examination was performed.

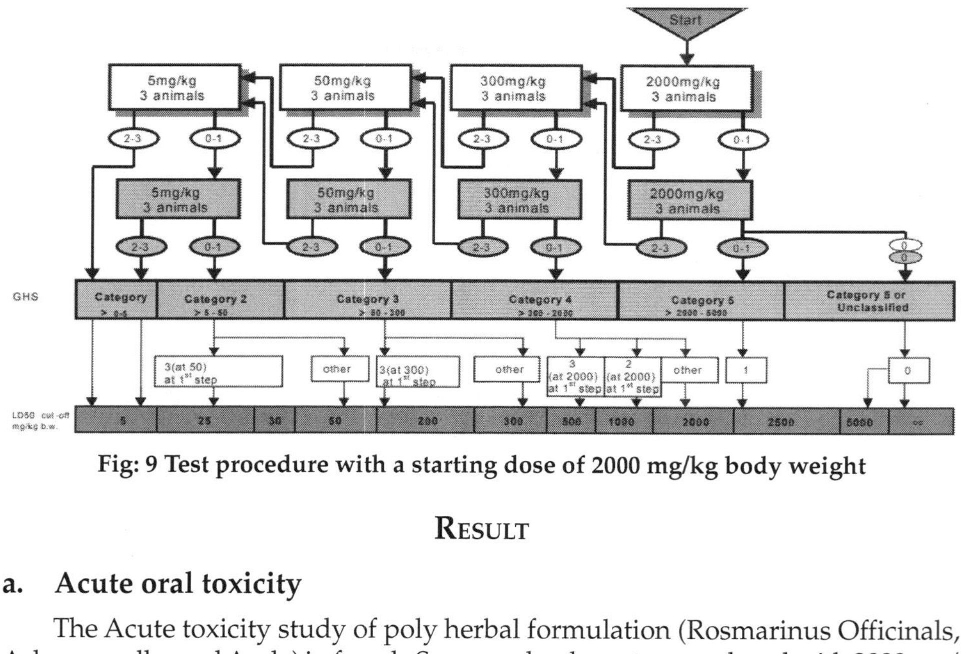

Fig: 9 Test procedure with a starting dose of 2000 mg/kg body weight

Result

a. Acute oral toxicity

The Acute toxicity study of poly herbal formulation (Rosmarinus Officinals, Ashwagandha and Amla) in female Sprague dawley rats were dosed with 2000mg/Kg had no adverse effect on the behavioural responses of the tested rats up to 14 days of observation. Physical observations indicated no signs of changes in the skin, fur, eyes mucous membrane, behaviour patterns, tremors, salivation, and diarrhoea of the rats. There was no mortality was observed and No body weight changes or anatomical abnormalities were noted at necropsy. Therefore, poly herbal extracts have a low risk of acute toxicity, and the oral lethal dosage (LD50) for female rats is higher than 2,000 mg/kg of body weight. It is very crucial to evaluate acute toxicity signs in attempt to establish a lethal dose causing of 50% of animals death (LD50). According to these findings, it was assumed that poly herbal formulation LD50 dose is above 2000 mg/kg.

b. Body weight

There were no treatment related changes in body weight in any of the treated animals and showed gain on day 8 and 15, as compared to day 1 body weight. The Individual animal body weights (g) are presented in Table-1 showed the change observed before and after the administration of the poly herbal extracts.and food and water consumption thus were unaffected by administration of poly herbal extracts as a single oral dose of 2,000 mg/kg of body weight.

c. Clinical signs

In Step I (2000 mg/kg body weight), no clinical signs of toxicity was observed in any of the treated animals during Day 1 at 30 min, 1st hour, 2nd hour, 3rd hour and 4th hour after treatment and throughout experimental period of 15 days. In Step I (2000 mg/kg body weight), no clinical signs of toxicity was observed in any of the treated animals during Day 1 at 30 min, 1st hour, 2nd hour, 3rd hour and 4th hour after treatment and throughout experimental period of 15 days. The detail of clinical signs were observed.

d. Gross pathology

All animals were sacrificed at the end of the study and they did not reveal any abnormality of pathological significance. External examination of terminally sacrificed rats also did not reveal any abnormality of pathological significance. On the day of necropsy (Day 15), no external and internal macroscopical lesions were observed in gross pathological examination. The detail of individual gross pathological findings were observed. (Based on these findings of the acute oral toxicity (Acute Toxic Class Method) of the poly herbal formulation in Sprague dawley rats, the LD_{50} of the extract may be classified as GHS (Globally Harmonized System) category 5 (LD_{50} >2000 mg/kg bw) as per OECD Guideline No. 423, December 2001) [30].

Table: 1 INDIVIDUAL ANIMAL BODY WEIGHT (G)

Key: F: Female, SD – Standard Deviation

Step	Dose (mg/ kg body weight)	Sex	Animal No.	Volume Administered (ml)	Time of Dosing	Body Weight (gram) Days			Body Weight Gain (%)	
						Day 1	Day 8	Day 15	Day 1-8	Day 1-15
I	2000	F	1	1.7	10:20 to 10:25	171.39	187.32	203.05	9.29	18.47
			2	1.8		189.73	200.65	208.87	5.76	10.09
			3	1.9		185.99	202.18	218.7	8.70	17.59
			Mean			**179.06**	**195.38**	**182.37**	**196.72**	**210.21**
			SD			**7.37**	**9.06**	**9.69**	**8.17**	**7.91**

II	2000	F	4	1.8	10:18 to 10:23	179.9	197.58	213.91	9.83	18.90
			5	1.8		183.06	200.11	215.32	9.31	17.62
			6	1.9		187.89	204.15	222.44	8.65	18.39
			Mean			**183.62**	**200.61**	**183.62**	**200.61**	**217.22**
			SD			**4.02**	**3.31**	**4.02**	**3.31**	**4.57**

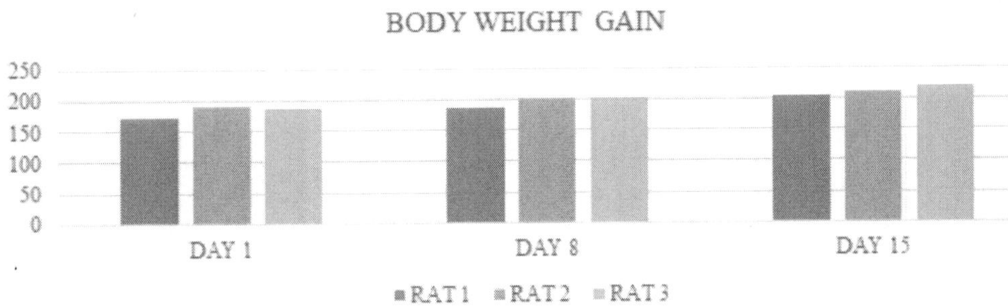

STEP-I: The Graph showing Body weight in Grams V/S Experimental period in Days.

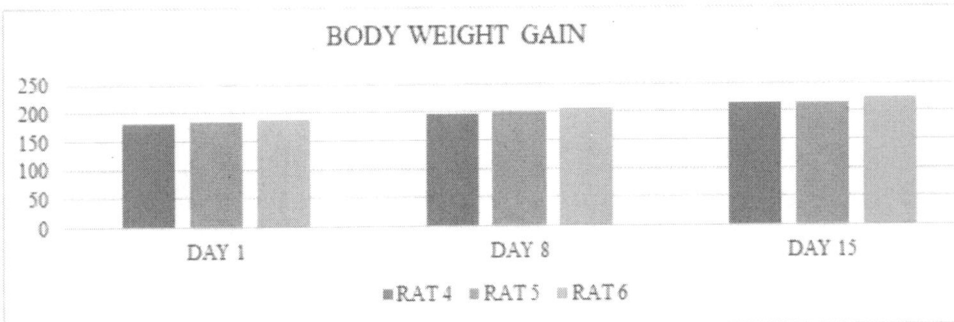

STEP-II: The Graph showing Body weight in Grams V/S Experimental period in Days.

Discussion

In order to assess the acute toxicity of polyhebal formulation (rosemerry, ashwaghandha and amla) in Sprague dawley rat, we applied OECD Guideline 423, which comprises a single dose and a 14-day observation period (OECD Guideline 423, 2001) [30]. Our findings indicate that polyhebal formulation was not acutely toxic to female Sprague dawley rats according to the Globally Harmonized System (GHS) for the classification of chemicals (OECD Guideline 423), with an LD_{50} cut-off value of 2000 mg/kg b.w. Medicinal plants are been used since centuries to treat different ailments. Hence, this project was designed to check the toxic effect of the plant by oral route using 425 toxicity guidelines. Toxicological assessments in experimental animals usually were categorized into four classes: acute, subacute, subchronic and chronic. Acute toxicity test is defined as a single exposure for less than 24 h, subacute toxicity refers to repeated exposures for 1 month or less, subchronic toxicity refers to repeated exposures for 1 to 3 months and chronic toxicity assay refers to repeated exposures for more than 3 months.24 The oral route of drug administration is the most convenient and commonly used method for toxicity evaluations in pre-clinical animal models. Acute oral toxicity test is conducted at a limit dose (2000mg/kg) primarily in situations where the experimenter has information indicating that the test material is likely to be non-toxic (OECD, 1998) [31]. The acute study was hence, conducted at the limit dose of 2000mg/kg and was found to be well tolerated. This maximum tolerated dose was taken as the highest dose for the sub-acute toxicity study. Two fold decreases were made from the highest dose to arrive the medium and low doses of the study, i.e. 1000 and 500 mg/kg body weight, respectively (OECD, 1995) [32]. Many official guidelines have been developed by various regulatory agencies for toxicity testing. These guidelines relate to the use of botanical products as medicinal preparations and provide standard methods for toxicological studies to assess the safety of medicinal products. Not all tests are necessarily performed for each herbal product and the need for each toxicity test should be evaluated depending on the availability of published literature related to safety, efficacy, established and traditional claims and intended uses [33]. 5 status (LD50> 5000mg/kg), which was the lowest toxicity class. According to the study by Kennedy et al. Substances with LD_{50} values higher than 5000mg/kg by oral route are regarded as being safe or practically nontoxic. However, many medicinal plants have also been reported to be toxic to both humans and animals. Therefore, it should be emphasized that the traditional use of any plant for medicinal purposes, by no means, guarantees the safety of such plant.

Toxicity studies are considered necessary, especially on drugs that are to be used in chronic conditions. About 80% of the world's people depend largely on traditional plant-derived drugs for their primary health care. Plant-derived drugs have an important place in both traditional and modern medicine. For this reason, a special effort to maintain the great diversity of plant species would undoubtedly help to alleviate human suffering in the long term. Medicinal plants constitute the base of health care systems in many societies. The drug under investigation herbal combination has been used for several years of its bio-stimulating, revitalizing and fertility enhancing effects by [34].There is, in recent times, a growing and increasing interest in herbal medicines. Consequently herbal medicines have received greater attention as an alternative to clinical therapy leading to increasing demand by [35]. Herbal medicines derived from plant extracts are being increasingly utilized to treat a wide variety of clinical disease. Even when efficient, the toxicity of the used preparations is usually unknown, and the population does not care, believing that if the preparation has been used so far, it should be devoid of toxicity by [36]. However, animal toxicity testing have shown many plants currently used, to name a few Momordica charantia by [37, 38, 39, 40, 41], as highly toxic when given either acutely or sub-chronically. The concept of polyherbalism has been highlighted in Sharangdhar *Samhita*, an Ayurvedic literature dating back to 1300 AD. From the long-term traditional folk use and clinical applications, it is fairly understood that most herbal plants or products derived from them have an excellent safety record [42, 43, 44].

The diverse therapeutic properties of *R. officinalis* has incremented its use by the population by [45, 46], however the presence of chemical compounds, such as fl avonoids, that have been reported to have toxic by [47, 48, 49] and antifertility activities by [50, 51] as well as the anti-implantation effect reported on female rats by [52] suggests a potential reproductive toxic role of this plant. Although rosemary extract did not produce acute toxicity, however, the European Commission has approved the internal use of rosemary leaf for dyspeptic complaints and the external use as supportive therapy for rheumatic diseases and circulatory problems. Because of its anti-inflammatory and antioxidant properties, rosemary is used to treat localized injuries. Withania *somnifera* is a vital medicinal plant traditionally utilized in the treatment of numerous diseases. Owing to the presence of some toxic alkaloids in the root powder and crude extracts, safety of the commercially available extracts has been raised. In few cases, it has been reported that ingestion of ashwagandha might cause stomach upset, diarrhea, and vomiting. The alkaloid rich part of commercial ashwagandha extract was found to be toxic in a zebra fish embryo acute toxicity test as per OECD guideline 236 [53] whereas detoxified ashwagandha extract (rich in glycowithanolides and containing very less alkaloids) was found to be non-toxic under the conditions of the study (under publication). Emblica *officinalis* prevents the retrain stress-induced oxidative stress and balancing the anti-oxidant system, since Emblica *officinalis* is reported as a rich source of Vitamin C [54]. In an acute oral toxicity study, it was observed that the lethal oral toxicity of the PHF was estimated to be higher than 2000 mg/kg, classified as category 5 according to OECD Guide 423, indicating a certain safety margin associated with the use of PHF as therapeutic agents [2]. The median acute toxicity value (LD_{50}) of the extract must be above 10ml /Kg body weight. According to [55, 56] the extract can be classified as non-toxic, since the LD_{50} was found to be more than 15.0 g/Kg.

It suggested that the normal processing of all the nutrients like carbohydrate, proteins and fats are been metabolised appropriately within the body as these are the nutrients that play a major role in physiological function. Postmortem toxicology of treated animals is customarily recommended in the adopted guidelines for acute toxicity testing by [57]. The gross pathological finding for each animal is genuinely considered as potential source of information on the target organ/system and the toxic nature of the chosen test substance. Necropsy examination conducted at the termination of 14 days observation study on individual animal did not show any significant treatment related macroscopic changes of organs or other structures. Brain, Liver, Kidney and heart are the major vital organs of the body that are been not affected by the toxic substance .When animals were sacrificed at the end of study, there were no lesions found on histological examination of Brain, heart, kidney and liver in comparison with vehicle control group. According to [58], substances with LD_{50} of 2000 mg/kg body weight (oral route) are regarded as being safe or of low toxicity.

The results of acute toxicity assay showed that "Polyherbal Formulation (Rosmarinus officinals, Ashwagandha and Amla)" at dose of 2000 mg/kg did not cause death and behavioural changes in the animals. Therefore, it can be concluded that according to OECD guidance, the poly herbal extract may be assigned to be the lowest toxicity class 5 (LD_{50}> 2000 mg/kg). In conclusion, acute oral toxicity testing of screened poly herbal products did not produce mortality, toxicity signs or any significant pathological changes on the dose level of 2000 mg/kg body weight and an overall normal body weight gain was observed in all the treated female rats and hence resulted in tested products being labelled unclassified in the hazard category according to Globally Harmonized System.

Conclusion

Polyherbal formulation treating for various disease bring attention to the human society. The toxicity and its safe dosage should be studied for dose optimization and formulation. The aim of this study was to evaluate the oral toxic effect of the polyherbal formulation of Ashwagandha, Rosemary and Amla on female Sprague dawley rats. The acute oral toxicity was evaluated as per OECD (Organization for Economic Co-operation and Development) guidelines 423. The animals will be acclimatized for a minimum period of five days to laboratory conditions. Veterinary examination will be performed before selecting the animals and healthy and active animals will be used in the study. Acute oral toxicity was tested by single dose oral administration of the drug at dose of 2000mg/kg body weight. The animals were identified by tail marking and using cage cards. The animals were observed daily for clinical signs of abnormality/ mortality for 14 days. After 14 days, the animals were weighed before necropsy. Body weight changes will be calculated and recorded. All the animals will be sacrificed by using carbon dioxide asphyxiation and subjected to gross necropsy examination. Microscopic examinations, haematological, biological parameters of treated rats were compared with control animals. Histopathology of all the major organs was also studied. In the present oral toxicity study of polyherbal formulation, the result shows no mortality, no significant clinical signs of toxicity were observed in any of the animals at maximum recommended dose level of 2000 mg/ kg body weight. No treatment related gross/histopathological lesions were observed. The haematology and biochemistry profile of treated rats was similar to control animals and difference was non-significant. The histopathology of major organs of all control animals was normal. In this study the NOAEL (No Observed Adverse Effect Level) was calculated as 1000 mg/ kg bw for rats.

The present study clearly indicates that polyherbal formulation of Ashwagandha, Rosemary and Amla does not have any significant toxic effects in animals at the dose evaluated as oral acute toxicity studies in female Sprague dawley rats. The present study concludes that the oral toxicity results of polyherbal formulation of Ashwagandha, Rosemary and Amla and their individual components emphasized to considered it an no toxic effect to the rats. This study also validated the traditional use of natural remedies as indigenous plants origin for the treatment of diseases. The active constituents present in the Ashwagandha, Rosemary and Amla improves their antioxidant capacity and anti-inflammatory property. So, the present study demonstrated that the polyherbal formulation of Ashwagandha, Rosemary and Amla to be safe without significant toxicity, as it neither caused mortality nor produced any significant haematological, biochemical, gross and histopathological changes, in female Sprague dawley rats. Since, the oral dose of 2000mg/kg bw of polyherbal formulation was the highest dose (limit dose) in an acute oral studies and that it did not cause any adverse effects, it is concluded as the no observed-adverse-effect level for female Sprague dawley rats under the experimental conditions used. Further, studies have to be undertaken to assess the other aspects of toxicities like long term toxicity (90 days/1 year), reproductive toxicity, teratogenicity and carcinogenicity.

Acknowledgement

I would like to thank my beloved parents, guide, colleagues and my dear students for their assistance and guidance during my work and I would also like to thank our beloved Management, Principal and HOD, Department of Zoology, LVD College Raichur, Karnataka, India for their support and facility provided during my work.

References

1. Pandey G, Pandey SP. Medicinal plants against liver diseases. IJPR. 2011; 2(5):115-121.
2. Belhekar SN, Chaudari PD. Acute and sub-acute oral toxicity assessment of the polyherbal formulation in albino Wistar rats. Int Pharm Sci. 2016; (8):311-316.

3. Trevan JW. The Error of Determination of Toxicity, Proc. R. Soc. Lond. B. 1927; (101):483-514.
4. OECD. Test Guideline 401: Acute Oral Toxicity. OECD Guidelines for Testing of Chemicals. 1981.
5. Thomas LF, Landsiedel R, Lebreux F. Integrated Testing Strategies (ITS) for safety assessment. ALTEX. 2015; (1):32.
6. Asima C, Satyesh CP. The treatise on Indian medicinal plants, New Delhi: Publication and Information Directorate. 1995; (3).
7. Anderson D, Cheng Y, Duan RD. Ursolic acid inhibits the formation of aberrant crypt foci and affects colonic sphingomyelin hydrolyzing enzymes in azoxymethane-treated rats. J. Cancer Res. Clin. Oncol. 2008; (134):101–107.
8. Susan C, Joseph T. Anti -proliferative and antioxidant properties of rosemary Rosmarimus officinalis. Oncology reports. 2007; 17(6):1525-1531.
9. Rajurker S, Rekhe DS, Maini S, Ravikanth K. Acute toxicity studies of polyherbal formulation (Methiorep premix). Veterinary world. 2009; 2(2):58.
10. Michael R. Herbal medicine: expanded commission E monographs. Annals of internal medicine. 2000; 133(6):487-487.
11. Al-Sereiti MR, Abu-Amer KM, Sen P. Pharmacology of rosemary (Rosmarinus officinalis Linn.) and its therapeutic potentials. Indian J. Exp. Biol. 1999; 37(2):124-130.
12. Asiya B, Subarda S, Kombath RV, Swapna R, David B. An in -depth review on the medicinal flora Rosmarinus officinalis. (Lamiaceae). Acta scientiarum polonorum technologia alimentaria. 2013; 12(1):61-74.
13. Comissˆao Permanente da Farmacopeia Portuguesa. Farmacopeia Portuguesa (7th Edition). Minist´erio da Sa´ude, Lisbon, Portugal. 2003.
14. Aruoma OI, Halliwell B, Aeschbach R, Löligers J. Antioxidant and pro-oxidant properties of active rosemary constituents: carnosol and carnosic acid. Xenobiotica. 1992; 22(2):257-268.
15. Changhadi GS. Ashwagandharishta—Rastantra Sar Evam Sidhyaprayog Sangrah. Krishna-Gopal Ayurveda Bhawan (Dharmarth Trust), Nagpur. 1938; 743-774.
16. Sivarajan VV, Balachandran I. Ayurvedic drugs and their plant sources. International Science Publisher, New York. 1994.
17. Warrier PK, Nambiar VPK, Ramankutty C. Indian Medicinal Plants: A Compendium of 500 Species. Orient Longman, Madras, India. 1996.

18. Usmanghani K, Saeed A, Alam MT. Indusyunic medicine: traditional medicine of herbal, animal, and mineral origin in Pakistan. Dept. of Pharmacognosy, Faculty of Pharmacy, University of Karachi, Karachi, Pakistan. 1997.

19. Kulkarni SK, Dhir A. Withania somnifera: an Indian ginseng. Prog. Neuropsychopharmacol. Biol. Psychiatry. 2008; (32):1093–1105.

20. Mishra LC, Singh BB, Dagenais S. Scientific basis for the therapeutic use of Withania somnifera. (Ashwagandha): A review. Alternative Medicine Reviews. 2000; 5(4):334-346.

21. Bhattacharya SK, Goel RK, Kaur R, Ghosal S. Anti stress activity of sitoindosides VII and VIII, new acylsterylglucosides from Withania somnifera. Phytotherapy research. 1987; 1(1):32-37.

22. Ghosal S, Srivastava RS, Bhattacharya SK, Upadhyay SN, Jaiswal AK, Chattopadhyay U. Immunomodulatory and CNS effects of sitoindosides IX and X, two new glycowithanolides form Withania somnifera. Phytother. Res. 1989; (2):201– 206.

23. Atta-ur-Rahman, Samina-Abbas, Dur-e-Shahwar, Jamal SA, Choudhary MI, Abbas S. New withanolides from Withania spp. Journal of Natural Products. 1991; (56):1000–1006.

24. Shweta Khosla, Sunny sharma. A short description on pharmacogenetic properties of Emblica officinalis. Spatula DD. 2012; 2(3):187- 193.

25. Swetha D, Krishna MG et. al. Current trends in the research of Emblica officinalis (Amla) : A pharmacological perspective. Int J Pharm Sci Rev. 2014; 24(2):150-159.

26. Arunabh B, Shibnath G, Sahil KB. Antioxidant activity of tannoid principles of Emblica officinalis (amla) in chronic stress induced changes in rat brain. NISCAIR- CSIR, India, 2000.

27. David JV, Thompson MD, Cook EH, Bendahhou E. Human dopamine transporter gene: coding region conservation among normal, Tourette's disorder, alcohol dependence and attention- deficit hyperactivity disorder populations. Molecular psychiatry. 2000; 5(3):283- 292.

28. AAALAC. The evolution and adoption of standards used by AAALAC. Journal of the American Association for Laboratory Animal Science. 2012; 51(3):293-297.

29. CPCSEA. Animal experimentation and ethics in India: the CPCSEA makes a difference. Alternatives to laboratory animals. 2004; (32):411-415.

30. OECD. Guideline for the Testing of Chemicals. Acute Oral Toxicity e Acute Toxic Class Method: Test No-423. Organization for Economic Cooperation and Development. 2001.

31. OECD. An Emerging Global Issue (1998-04-09) [2015-11-28] http://www.oecd.org/tax/transparency/44430243.pdf. 1998.

32. OECD. Repeated dose 28-day oral toxicity test method guideline 407 adopted 27.07.1995. In: OECD, Guidelines for Testing of Chemicals. Organization for Economic Cooperation and Development, Paris. 1995.

33. Schilter B, Andersson C, Anton R, Constable A, Kleiner, J et. al. Guidance for the safety assessment of botanicals and botanical preparations for use in food and food supplements. Food and Chemical Toxicology. 2003; 41(12):1625-1649.

34. Gauthaman K, Adaikan PG, Prasad RNV. Aphrodisiac properties of Tribulus Terrestris extract (Protodioscin) in normal and castrated rats. Life sciences. 2002; 71(12):1385- 1396.

35. Mythilypriya R, Shanthi P, Sachdanandan P. Oral acute and subacute toxicity studies with Kalpaamurthaa, a modified indigenous preparation, on rats. Journal of Health science. 2007; 53(4):351-358.

36. Albert DA, Telesphore B, Nguelefack JY, Datte AK. Acute and subchronic oral toxicity assessment of the aqueous extract from the stem bark of Erythrin senegalis DC (Fabaceae) in rodents. Journal of Ethnopharmacology. 2011; 134:697-702.

37. Raman, Lau. Anti - diabetic properties and phytochemistry of Momordica charantia L. (Cucurbitaceae). Phytomedicine. 1996; 2(4):349- 362.

38. Ethan Basch, Steven Gabardi, Catherine ulbricht. Bitter melon (Momocordica charantia): A review of efficacy and safety. American Journal of Health-System Pharmacy. 2003; 60(4):356-359.

39. Tahri A, Yamani S, Lagssyer A, Mohammed A et. al. Acute diuretic, natriuretic and hypotensive effects of a continuous perfusion of aqueous extract of Uratica dioica in the rat. Journal of Ethnopharmacology. 2000; 73(1-2):95- 100.

40. Hosseinzadesh, Younesi. Antinociceptive and anti-inflammatory effects of Crocus sativus L. stigma and petal extracts in mice. BMC pharmacology.2002; 2(1):7.

41. Adeoye BA, Oyedapa OO. Toxicity of Erythrophleum guineese stem-bark: role of alkaloidal fraction. African Journal of Traditional Complementary and Alternative Medicine. 2004; (1):45 -54.

42. Bhattacharjee SK. Suggestions for Improvement of Medicinal Crop Industry, Handbook of medicinal plants.1st Edn,Pointer Publishers, Jaipur. ISBN: 81-7132-156-9, 1998; 384.

43. Agaie BM, Onyeyili PA, Muhammad BY, Ladan MJ. Acute toxicity effects of the aqueous leaf extract of Anogessus leiocarpus in rats. African Journal of Biotechnology. 2007; 6(7).

44. Michael H. Herbal medicines in health care- Benefits and risks. Toxicology, 2007; 240:129-129.

45. Ruppelt BM, Pereria EFR, Goncalves LC, pereria NA. Preliminary assessment of Rosmary officinalis toxicity on male Wistar rats' organs and reproductive system. Revista Brasilereira de Farmacognosia. 2006; 16(3):324-332.

46. Cruz FG, Roque NF, Giesbrecht AM, Davino SC. Antibiotic activity of diterpenes from Mikania triangularis. Fitoterapia. 1996; 67:180-190.

47. Bhargava SK. Antiandrogenic effects of a flavonoid rich fraction of Vitex negundo seeds: a hostological and biochemical study in dogs. J Ethnopharmacol. 1989; 27:327-339.

48. Ulubelen A, Ertugrul L, Birman H, Yigit R, et. al. Antifertility effects of some coumarins isolated from Ruta chalepensis and R. chalepensis var. latifolia in rodents. Phytotherapy Research. 1994; 8(4):233-236.

49. Born SL, Caudill D, Smith BJ, Lehman-Mckeeman LD. In vitro kinetics of coumarin 3, 4-epoxidation application to species differences in toxicity and carcinogenecity. Toxicol Sci. 2000; 58:23-31.

50. Vilegas JHY, Marchi E, Lancas FM. Extraction of low-polarity compounds (with emphasis on coumarin and kaurenoic acid) from Mikania glomerata ("guaco") leaves. Phytochem Anal. 1997; (8):266-270.

51. Martins ER, Castro DM, Castellani DC, Dias JE. Plantas Medicinais. Viçosa: Editora UFV, 2000; 70-73.

52. Lemonica IP, Damasceno DC, Di-Stasi LC. Study of the embryotoxic effects of an extract of rosemary (Rosmarinus officinalis L.). Brazilian journal of medical and biological research= Revista brasileira de pesquisas medicas e biologicas. 1996; 29(2):223-227.

53. OECD guidelines 236. Applicability of the fish embryo acute toxicity (FET) test (OECD 236) in the regulatory context of Registration, Evaluation, Authorisation, and Restriction of Chemicals (REACH).

54. Settu Dinesh Kumar, Mugasaparur Ganesan Rajanandh, Chamundeeswari Duraipandian. Acute and Sub-acute Toxicity Studies of a Patented Anti-anxiety Poly Herbal Formulation. Toxicology. 2019; 14(1):9-17.

55. Ghosh MN. Fundamentals of Experimental Pharmacology, 2nd Edition. Scientific Book Agency, Calcutta. 1984; 154-157.

56. Klaasen CD, Amdur MO, Doull J. Casarett and Doull's Toxicology: The basic science of poison. 8th Edition. Mc Graw Hill, USA. 1995; 13-33.

57. Dadarkar SS, Deore MD, Gatne MM. Comparative evaluation of acute toxicity of ivermectin by two methods after single subcutaneous administration in rats. Regulatory Toxicology and Pharmacology. 2007; 47(3):257-260.

58. Clarke EGC, Clarke ML. London: Cassel and Collier Macmilan. Veterinary toxicology. 1977; 268–277.

CHAPTER 5

STUDY ON THE DIVERSITY OF DIFFERENT GROUPS OF INSECTS IN AND AROUND DAVANGERE UNIVERSITY CAMPUS DAVANGERE

Ms. Reshma M. D.

**Department of studies in Zoology, Davangere University, Davangere-577007*

Abstract:

The insect diversity of Davangere University was carried out under the aegis of DUSMYTR seed money grant to identify different insect orders in the study area from July 2021 - October 2021 by employing different survey techniques. The study recorded the presence of insects belonging to the 10 insect enlisted ahead; Blattodea, Coleoptera, Diptera, Hemiptera, Hymenoptera, Lepidoptera, Mantodea, Odonata, Orthoptera and Thysanura. Among them order Lepidoptera was the dominant order with respect to diversity. The study indicates the presence of a good number and diversity of species in and around Davangere University campus. However, further detailed, exhaustive, systematic investigation of the insect fauna in the study area is needed to be conducted to update the checklist.

Keywords: Davangere, Diversity, Insects, Fauna, Biodiversity

INTRODUCTION

Insects are the only invertebrates with wings. They belong to the phylum arthropoda and class Insecta. Their distinguishing morphological features include the presence of hard exoskeleton, clearly visible head, thorax, abdomen, two pairs of wings and three pairs of legs. They inhabit almost all natural, modified habitats and are found practically everywhere. They are considered as extremely successful animals and they affect many aspects of our lives despite of their small size.

The class insecta will constitute the dominant group of animal kingdom representing 80 to 92% of the entire animal kingdom. They are one of the dominant groups of organisms with respect to their numbers and diversity. More than 10, 50,000 species of insects have been described worldwide till today, yet there are millions of kinds of insects that are yet to be discovered and described.

Insects play an important role in terrestrial ecosystems as herbivorous, pollinators, decomposers, predators and parasites. They provide environmental benefits by delivering several ecosystem services like nutrient cycling via leaf litter, wood degradation, disposal of dung and carrion, dispersal of pollens and seeds, soil turnover, prey for insectivorous organisms and predators for some organisms thereby contributing to the food chains and food webs. They are the major contributors of macroscopic biodiversity and hence their study is highly essential. Apart from this some insect groups provide economic benefits like silk moths, honey bees, lac insects and several edible insects. Several insects are used in solving medico-criminal investigations.

The effects of insects are enormous. As some help in pollination service some cause damages to the crops and livestock, thereby causing economic loss. Some insects act as parasites and vectors may aid in transmitting diseases. Thus they affect the health of organisms including humans. Yet some of them can be used to control the pests' population. Characteristics such as the short generation time, high fecundity rate, ease of laboratory rearing and manipulation of certain insects make them useful models for understanding general biological processes for instance Drosophila melanogaster.

Despite the benefits and threats posed by insects, some of them are sensitive to the rapid changes in the environment. Some species can adapt and others can struggle to survive. Due to use of various chemicals in the agriculture and other sectors there has been an observed decline in the insect population. Hence there is a need to explore the insect diversity to identify the insects inhabiting a particular area as the knowledge about their habitat, diversity, population, life cycle and behavior is of significance.

Davangere university campus situated 10 KM away from the main city Davangere has pollution free campus with wide area of 76 acres. Many regions of the campus have less human interference except for the academic and administrative buildings and surrounding areas. The campus has rich tree cover with respect to both number and diversity. Further, there are aquatic resources and many micro and macro habitats are available within and around the campus. Field observations reveal the presence of rich diversity of birds and some groups of insects in the campus. But as the information available about the diversity of insects is limited hence this work is proposed to conduct preliminary study to identify the diversity of different insect groups in and around the shivagangothri campus.

Materials and Methods

Study Area

Davangere University campus is situated near Tholahunase village about 10 kilometres away from the city. The campus covers an area of 73 acres and is surrounded by agricultural fields, village premises, open fields and state highway. The study area is predominantly covered by well-maintained garden areas, shrubby region, teak trees, and many fruit bearing plants, trees. It also has open fields, large buildings and small aquatic bodies. There are regions where more human interference can be seen like the regions surrounding academic and administrative blocks and hostels. There are some regions where moderate to less human interference is observed like the teak plantation regions, shrubby vegetation regions and garden areas.

Fig: 1 Map of the study area: Davangere District and Davangere University campus respectively.

Sampling methods

The survey was carried out from July 2021 to October 2021. Surveys were conducted once in a week in and around the campus to cover all the habitats. To observe the nocturnal insects light traps were set in different locations within the campus. Only adult fauna was recorded during the study by employing direct searching and observation method at the potential habitats of insects. For this purpose, 10–15 min halts were made at each search point chosen at random. Opportunistic sightings were also recorded. Most of the insects were observed, captured, identified and released immediately at the spot of capture. Individual images of them were photo-documented. Some insects which were difficult to identify in the field were collected using an insect net, hand picking method for proper identification of the specimen. The dead and damaged insects observed during the field visits were preserved following either dry or wet preservation methods. Their photographs were documented by using Canon power shot and Nikon D5600 cameras. The collected insects were identified using identification keys provided by standard identification manual, field guides and photo guides.

Results and Discussion

Identification and enumeration the diversity of insects in and around the shivagangothri campus:

The field observations during the study reveal the presence of different groups of insects in and around Davangere University campus. Insects belonging to the following 10 insect orders were recorded in the study area from July 2021 - October 2021, they are; Blattodea, Coleoptera, Diptera, Hemiptera, Hymenoptera, Lepidoptera, Mantodea, Odonata, Orthoptera and Thysanura. Among them order Lepidoptera was the dominant with respect to diversity. A total of 62 species of moths and 31 species of butterflies belonging to lepidoptera were observed during the study. Least number of species were observed from order Thysanura with one species. A total of 17 species of odonates, 17 species of Hymenoptera, 7 species of grasshoppers, 5 species each from Diptera, Hemiptera and coleoptera, 4 species of mantodea and 2 species of blattodea were observed in and around Davangere University campus during the study.

Dead specimen that were collected during the survey were preserved using ethanol following either dry or wet preservation methods. Other insects that were not collected for preservation were photo documented. Most of the specimen were identified up to genus level and some up to species level using the appropriate literature, field and photo guides.

Table 1. Checklist of odonates recorded in the study area during the present study.

SL No.	Family	Common Name	Scientific Name
1	Aeshnidae	Blue-tailed Green Darner	*Anax guttatus*
2	Libellulidae	Ditch jewel	*Brachythemis contaminata*
3	Libellulidae	Ground skimmer	*Diplacodes trivialis*
4	Libellulidae	Crimson Tailed Marsh Hawk	*Orthetrum pruinosum*
5	Libellulidae	Crimson Marsh Glider	*Trithemis aurora*
6	Libellulidae	Emerald-Banded skimmer	*Cratilla lineata*
7	Libellulidae	Ruddy Marsh skimmer	*Crocothemis servilia*
8	Libellulidae	Rufous marsh glider	*Rhodothemis rufa*
9	Libellulidae	Slender Skimmer	*Orthetrum sabina*
10	Libellulidae	Common picture wing	*Rhyothemis variegata*
11	Libellulidae	Orange-winged drop wing	*Trithemis kirbyi*
12	Libellulidae	Wandering Glider	*Pantala flavescens*
13	Coenagrionidae	Golden Dartlet	*Ischnura rubillio*
14	Coenagrionidae	Senegal golden dartlet	*Ischnura senegalensis*
15	Coenagrionidae	Rusty Marsh Darts	*Ceriagrion olivaceum*
16	Coenagrionidae	Blue Grass Dartlet	*Pseudagrion microcephalum*
17	Coenagrionidae	Coromandel Marsh Dart	*Ceriagrion coromandelianum*
Total	**03**	**17s**	**17**

Table 2. Checklist of moths recorded in the study area during the present study.

SL No.	Family	Common Name	Scientific Name
1	Bombycidae	-	*Trilocha varians*
2	Crambidae	-	*Bradina sps.*
3	Crambidae	Cucumber moth	*Diphonia indica*
4	Crambidae	-	*Endocrisis flavibasalis*
5	Crambidae	Bicolored pearl	*Glypodes bicolr*
6	Crambidae	Fruit and shoot borer	*Leucinodes orbanalis*
7	Crambidae	Banded pearl	*Sameodes cancellasis*
8	Crambidae	-	*Hydridis ornatalis*

9	Crambidae	-	*Prophantis adusta*
10	Crambidae	Waterlily leafcutter moth	*Elophila obliteralis*
11	Crambidae	Beet webworm moth	*Spoladea recurvalis*
12	Crambidae	Lima bean pod borer	*Maruca vitrata*
13	Erabidae	-	*Amata passalis*
14	Erabidae	Trimen's knob	*Acantholipes trimeni*
15	Erabidae	-	*Brunia antica*
16	Erabidae	-	*Creatonotos gangis interrupta complex*
17	Erabidae	-	*Cyana pergrina*
18	Erabidae	Common fruit-percing moth	*Endocima phalonia*
19	Erabidae	-	*Lacera noctillo*
20	Erabidae	-	*Olepa ocellifera*
21	Erabidae	-	*Dichromia sagitta*
22	Erabidae	Common owl moth	*Erebus macrops*
23	Erabidae	Brown tussock moth	*Olena mendosa*
24	Erabidae	Common Nyctemera	*Nyctemera lacticinia*
25	Erabidae	Semi-looper moth	*Trigonodes hyppasia*
26	Erabidae	Nygmiine tussock moth	*Orvasca subnotata*
27	Erabidae	-	*Lymantria marginate*
28	Erabidae	Handmaiden moth	*Syntomoides imaon*
29	Erabidae	Heliotrope moth	*Utetheisa pulchelloides*
30	Erabidae	Red coasted tigermoth	*Aloa lactinia*
31	Erabidae	Footman moth	*Nepita coniferata*
32	Erabidae	-	*Hypopyra vispertillo*
33	Erabidae	-	*Ammatho cuneonotatus*
34	Erabidae	-	*Hypena sps.*
35	Erabidae	Brown striped semilooper moth	*Mocis undata*
36	Erabidae	-	*Hulodes caranea*
37	Erabidae	Sandalwood defoliator	*Anomis Combinans*
38	Erabidae	Sugarcane looper	*Mocis frugalis*
39	Erabidae	Jute hairy caterpillar	*Spilosoma obliqua*
40	Erabidae	-	*Padenia transversa*
41	Erabidae	-	*Chilkara fulcata*
42	Geometridae	-	*Hyperythra lutea*

43	Geometridae	-	*Pelagodes falsaria*
44	Geometridae	-	*Petelia medardaria*
45	Geometridae	-	*Eumelea*
46	Noctuoidae	Olive shaped bird drappping moth	*Ponometia candefacta*
47	Noctuoidae	-	*Ischya manlia*
48	Noctuoidae	Triangular striped moth	*Chalciope mugdon*
49	Noctuoidae	-	*Maliattha signifera*
50	Noctuoidae	Slender burnished brass	*Thysanoplusia orichaiceea*
51	Noctuoidae	-	*Anticarsia irrorata*
52	Nolidae	Tree mallow	*Xanthopodes albugo*
53	Notodontidae	-	*Allata sps.*
54	Pterophoridae	-	*Pterophoridae*
55	Saturnidae	-	*Antherara palpia*
56	Sphingidae	Tusar silkmoth	*Theretra silhetensis*
57	Sphingidae	Dark striated hawkmoth	*Hippotion velox*
58	Sphingidae	-	*Psilogramma vates*
59	Sphingidae	Swinshoes striated hawkmoth	*Hippotion rosetta*
60	Sphingidae	Palebrown hawkmoth	*Theretra Laterillii*
61	Thyrididae	-	*Striglina scitaria*
62	Uraniidae	-	*Micronia aculeata*
Total	**12**	**62**	**62**

Table 3. Checklist of butterflies recorded in the study area during the present study.

SL No.	Family	Common Name	Scientific Name
1	Lycaenidae	Common ciliate blue	*Anthem emolus*
2	Lycaenidae	Common pierrot	*Castalius rosimon*
3	Lycaenidae	Club silver line	*Cigaritis syama*
4	Lycaenidae	Dark grass blue	*Zizeeria karsandra*
5	Lycaenidae	Forget me not	*Catochrysopsstrabo*
6	Lycaenidae	Indian sub beam	*Curetis thetis*
7	Lycaenidae	Pale grass blue	*Pseudozizeeria maha*
8	Lycaenidae	Pea blue	*Lampides boeticus*
9	Lycaenidae	Zebra blue	*Leptotes plinius*
10	Nymphalidae	Angled caster	*Ariadne ariadne*

11	Nymphalidae	Common buckeye	*Junonia coenia*
12	Nymphalidae	Common crow	*Euploea core*
13	Nymphalidae	Common evening brown	*Melanitis leda*
14	Nymphalidae	Common palmfly	*Elymnias hypermnestra*
15	Nymphalidae	Common purring	*Hpthima huebneri*
16	Nymphalidae	Common three ring	*Ypthima asterope*
17	Nymphalidae	Creamy sailer	*Neptis soma*
18	Nymphalidae	Danaid eggly	*Hypolimnas msippuse*
19	Nymphalidae	Double branded crow	*Euploea sylvester*
20	Nymphalidae	Lemon pansy	*Junonia lemonias*
21	Nymphalidae	Lilacine bush brown	*Mycalesis francisca*
22	Nymphalidae	Long brand bush brown	*Mycalesis visala*
23	Nymphalidae	Peacock pansy	*Junonia almana*
24	Nymphalidae	Plane tiger	*Danus chrysippus*
25	Nymphalidae	Tawny caster	*Acraea terpsicore*
26	Papilionidae	Crimson rose	*Pachliopta hector*
27	Papilionidae	Lime butterfly	*Papiliio demoleus*
28	Pieridae	Common grass yellow	*Eurema hecabe*
29	Pieridae	Mimosa yellow	*Pyrisitia nisa*
30	Pieridae	Motted emigrant	*Catopsilia pyranthe*
31	Pieridae	Small grass yellow	*Eurema brigitta*
Total	**04**	**31**	**31**

Table 4. Checklist of grasshoppers observed in the study area during the present study.

SL No.	Family	Common Name	Scientific Name
1	Acrididae	Short-horned grasshopper	*Acrida sps.*
2	Acrididae	Slanted faced grasshopper	*Acrida cineria*
3	Acrididae	Giant slant faced grasshopper	*Acrida sps.*
4	Acrididae	Silent slant faced grasshopper	*Oedaleus abruptus*
5	Gryllidae	Tropical house cricket	*Gryllodes sps.*
6	Gryllidae	Cricket	*Loxoblemmus sps.*
7	Trigonidiidae	Small cricket	*Dianemobius sps.*
Total	**03**	**07**	**07**

Table 5. Checklist of ants recorded in the study area during the present study.

SL No.	Sub-Family	Common Name	Scientific Name
1	Dolichoderinae	Reddish brown colored ants	Dolochoderus sps.
2	Formicinae	Long necked sugar ant	Componotus sps.
3	Formicinae	Giant honey ant	Componotus sps.
4	Formicinae	Golden backed ant	Componotus sps.
5	Formicinae	Weaver ant	*Oecophylla smaragdina*
6	Myrmicinae	Common broad acrobat ant	*Crematogaster sps.*
7	Myrmicinae	Pharaoh ant	*Monomorium sps.*
8	Myrmicinae	Short legged hunchback ant	*Myrmicaria brunnea*
9	Myrmicinae	Spiny Harvest ant	*Pheidole sps.*
10	Myrmicinae	Common red fire ant	*Solenopsis geminata*
11	Myrmicinae	Miniscule house ant	*Tetramorium smithi*
Total	03	11	**11**

Table 6. Checklist of other insects observed in the study area during the present study.

SL No.	Order	Common Name	Scientific Name
1	Blattodea	Bush cockroach	Blata sps.
2	Blattodea	American cockroach	Periplaneta americana
3	Coleoptera	Ladybird beetle	Cheilomenes sps.
4	Coleoptera	Ladybird beetle	Cheilomenes sps.
5	Coleoptera	Ladybird beetle	*Sticholotis sps.*
6	Coleoptera	Click beetle	*Agrypnus sps.*
7	Coleoptera	Click beetle	*Lanelater sps.*
8	Diptera	Housefly	*Musca domestica*
9	Diptera	Bluebottle fly	Chrysomya sps.
10	Diptera	Common fruit fly	*Drosophila sps.*
11	Diptera	Mosquito	*Aedes sps.*
12	Diptera	Mosquito	*Culex sps.*
13	Hemiptera	Assasin bug	*Oncocephalus sps.*
14	Hemiptera	Shield bug	Tolumnia sps.
15	Hemiptera	Jewel bug	*Chrysocoris sps.*
16	Hemiptera	Leaf bug	*Lygus sps.*
17	Hemiptera	Cotton bug	*Dysdercus sps.*
18	Hymenoptera	Tropical hornet	*Vespa tropica*
19	Hymenoptera	Potter wasp	Delta sps.
20	Hymenoptera	Wasp	*Rhynchium sps.*

21	Hymenoptera	Giant Honey bee	*Apis dorsata*
22	Hymenoptera	Small Honey bee	*Apis cerana*
23	Hymenoptera	Carpenter bee	*Xylocopa sps.*
24	Mantodea	Stick mantis	Schizocephala sps.
25	Mantodea	Praying mantis	*Hierodula sps.*
26	Mantodea	Mantis	*Statilia sps.*
27	Thysanura	Silverfish	*Lepisma sps.*
Total	**07**	**27**	**27**

Conclusion

From the present study it was noted that good number and diversity of insects were inhabiting the study area due to presence of diversity of micro and macro habitats, availability of host plants, food and water resources. The study provides a preliminary checklist of 10 insect orders observed in study area. However, this study is only indicating but not exhaustive as it is only a field based preliminary study. The study indicates that there is a good number and diversity of species in and around Davangere University campus. However, the present study also has some limitations due to time constraints. Further, detailed, exhaustive, systematic investigation of the insect fauna in the study area is needed to be conducted in near future so that the checklist can be updated regularly.

Acknowledgement

The author would like to express their deep sense of gratitude to the Internal Quality Assistance Cell (IQAC), Davangere University, Davangere-577007 for providing DUSMYTR -2021 Seed grant- for carrying out the present study. The author would like to express their gratitude to the Co-Ordinator, DoS in Zoology, Davangere University, and all the students, teaching and non-teaching staff for their direct and/or indirect support in the work.

Grant Support and Financial Assistance

The study was carried out with the support of DUSMYTR -2021 Seed grant provided by Internal Quality Assistance Cell (IQAC), Davangere University, Davangere-577007.

References

Alavandi, S., Mamatha.S.T., Pankaja, Sowmya, H.P.(2019).A preliminary observation and Diversity of Butterflies of Shivangangotri Campus, Davangere University, Davangere, India.International Journal of Agricultural Science and Research,9(4):273-280.

Ankalgi, S., and Jadesh, M. (2014). Diversity of dragonflies and damselflies (Odonata) from Ankalga village (Gulbarga district) Karnakata, India. International Journal of Recent Scientific Research, 5, (10): 1851-1853.

Ankalagi, S.,Jadesh.S.,(2014).Diversity of Butterflies from Ankalga Village Gulbarga District,Karnataka , India. International Journal of Rescent Scientific Research,5(6):1166-1169.

Balmer, E. (2007). A concise guide to butterflies and moths. First edition, Atlantic publishing, China. Pp:12-248.

Baruah, C., and Saikia P.K. (2015). Abundance and Diversity of Odonates in different habitats of Barpeta District, Assam, India. International Research Journal of Biological Sciences, 4, (9): 17- 27.

Basavarajappa, S., Krishna, V.G. and Santhosh, S. (2018). Butterflies species composition and diversity in a protected area of Nagarahole, Karnataka, India. International journal of biodiversity and conservation, 10 (10), 432-443.

Basumatary, P., Adhikary, D., Daimary, M., Basumatary, N., and Daimary, A. (2015). A Preliminary Study on the Diversity of Odonata in Bodoland University and its vicinity, Assam, India. International Journal of Scientific and Research Publications, 5, (6): 221-230.

Dhamorikar, A. (2016).A fieldguide to insects and spiders of Kanha tiger reserve. I Edition, resurrect publishers, India, Delhi. Pp:1-230.

Gurule, A.,and Nikam, M. (2013). Diversity of moths from northern, Maharastra , India.Journal of Threatened Taxa, 5,(12):4693-4713.

Harisha, M.N. (2016). Evaluation of Status and Diversity of Odonates of Kondajji Lake, Kondajji Village, Harihar Taluk, Davanagere District, Karnataka, India. Journal of Entomology and Zoology Studies, 4, (4): 384-388.

Harisha, M.N., and Hosetti, B.B. (2017). Status, Diversity and Conservation Threats of Odonates in Kundavada Lake, Davanagere District, Karnataka, India. Journal of Entomology and Zoology Studies, 5, (1): 312-316.

Kehimkar, I. (2016). BNHS field guide Butterflies of India, first edition, BNHS, Mumbai. Pp: 1-506.

Kumar, K.M., and Shetty,M. (2018). Diversity of insects from kundapura Taluq, Udupi district, Karnataka, India. International journal of Science and Research, 2319-7064.

Narendra, A., and Kumar, S. M. (2006). On a trail with ants- a handbook of ants of peninsular India, Tholasi prints India Pvt. Ltd, Bangalore. Pp:1-180.

Subramanian, K.A. (2009). Dragonflies of India- a fieldguide, II edition, Vigyan Prasar, DST, Noida.

CHAPTER 6

STUDY ON DIVERSITY OF BIRDS IN DAVANGERE DISTRICT, KARNATAKA

Girimallesh B, Gayathri, Hema, Kavya, Madhumathi, and Manglahouttigoudra

Zoology Department, D.R.M. Science College Davangere

ABSTRACT:

Birds are an essential component of ecosystems, contributing to seed dispersal, pollination, and pest control while also serving as indicators of environmental health. This study investigates the diversity and distribution of bird species in Davangere District, Karnataka, an area characterized by a variety of habitats, including wetlands, agricultural fields, forests, and urban landscapes. Field surveys were conducted across multiple locations, using direct observation, photographic documentation, and reference-based species identification to record avian diversity. A total of 25 bird species from multiple families were documented, highlighting the ecological richness of the region. The Shannon-Wiener Diversity Index (H' = 1.78) and Simpson's Diversity Index (D = 0.81) indicate moderate avian diversity, with species distribution varying across different habitat types. Birds such as Cattle Egrets, Kingfishers, and River Terns were predominantly found in wetlands, while species like Baya Weavers and Indian Rollers were more common in agricultural areas. Urban environments supported adaptable species such as House Sparrows, Pigeons, and Common Mynas. The study also identifies habitat loss, pollution, climate change, and agricultural expansion as key threats to bird populations in the district. The findings emphasize the need for habitat conservation, afforestation programs, and public awareness initiatives to protect avian biodiversity. This research serves as a baseline assessment of bird diversity in Davangere and underscores the importance of continued monitoring and conservation efforts to safeguard bird populations and their ecological roles.

Keywords: Davangere, Myna, Baya, Wetland and Pigeon

INTRODUCTION

Birds play a vital role in maintaining ecological stability, functioning as pollinators, seed dispersers, insect controllers, and bio indicators of environmental health (Gaston, 2010). As one of the most diverse groups of vertebrates, birds are distributed across a wide range of habitats, from dense forests and wetlands to urban landscapes (Sekercioglu, 2006). India is home to an estimated 1,314 bird species, with Karnataka being a significant contributor to this biodiversity (Praveen et al., 2016). The avifauna of Karnataka includes species found in a variety of ecosystems such as grasslands, wetlands, deciduous forests, and urban areas (Kumar & Jayapal, 2020). Davangere District, located in central Karnataka, offers a diverse ecological landscape that supports a wide range of bird species. The district comprises agricultural lands, water bodies, and urban settlements, providing nesting and feeding grounds for both resident and migratory birds (Ali & Ripley, 1983). However, increasing urbanization, deforestation, pollution, and climate change are posing threats to avian biodiversity (Bird Life International, 2019). Birds dependent on wetlands and agricultural fields are particularly vulnerable to habitat degradation and changes in land-use patterns (Mukherjee et al., 2018).

Studying bird diversity in Davangere is essential for understanding species composition, distribution, and habitat preferences. Such research provides valuable insights into the ecological health of the region and helps in formulating effective conservation strategies (Bibby et al., 2000). Additionally, regular avian monitoring can aid in detecting population trends, seasonal migrations, and behavioral adaptations in response to environmental changes (Sutherland et al., 2004). This study aims to document the bird species present in Davangere District, analyze their abundance and distribution, and assess the potential threats to their survival. The findings will contribute to bird conservation efforts by providing baseline data that can be used in future ecological assessments and biodiversity conservation initiatives.

MATERIALS AND METHODS

Study Area

The study was conducted in **Davangere District, Karnataka**, covering key locations such as:

1. **Kunduvada Lake**
2. **Kondajji Lake**

3. **Harihara Bathi**
4. **Devarabelakere**
5. **Hiremegalagere**
6. **Karur**

These sites were selected based on their diverse habitats, which include wetlands, forests, and cultivated fields.

Survey Methodology

1. Active Searching Metho

   Birds were actively observed for over an hour in each location, particularly near riverbanks, trees, fields, and urban parks.

2. Visual and Photographic Identification

   A Sony DSLR camera was used to capture images for species identification.

   Binoculars were employed to observe birds from a distance without disturbing their natural behavior.

3. Field Data Recording

   Observations were recorded in a notebook, including species count, behavior, habitat type, and time of sighting.

4. Species Identification

   Birds were identified using standard field guides such as The Book of Indian Birds by Salim Ali and Birds of the Indian Subcontinent by Grimmett, Inskipp & Inskipp

   Online databases like eBird and Avibase were used for cross-verification.

Statistical Analysis

To assess bird diversity in Davangere District, the following biodiversity indices were calculated:

1. Shannon-Wiener Diversity Index (H')

   Measures species diversity considering both richness (total species) and evenness (distribution of individuals among species).

   Formula: $H' = -\sum (p_i \ln p_i)$

   where **p_i** is the proportion of individuals of each species.

2. **Simpson's Diversity Index (D)**

   Evaluates species dominance, where a higher value indicates greater biodiversity.

   Formula: $D = 1 - \sum (p_i^2)$

3. Chi-Square Test

   Applied to determine if the observed species distribution significantly deviates from an expected even distribution.

Results and Discussion

A total of 25 bird species belonging to multiple avian families were documented during the study. The species composition included waterbirds, forest-dwelling birds, agricultural birds, and urban species, indicating a diverse avian assemblage in Davangere District. Some of the notable species observed include the Asian Koel (*Eudynamys scolopaceus*), Baya Weaver (*Ploceus philippinus*), Cattle Egret (*Bubulcus ibis*), House Sparrow (*Passer domesticus*), and Indian Peafowl (Pavo cristatus). The presence of species from various ecological niches highlights the district's ability to support a wide range of avian fauna. The Shannon-Wiener Diversity Index (H' = 1.78) indicates a moderate level of species diversity, suggesting that multiple species coexist in the study area without any single species dominating excessively (Magurran, 2004). The Simpson's Diversity Index (D = 0.81) further supports this, demonstrating a well-balanced bird community with relatively high species evenness. The Chi-square test for species distribution revealed a statistically significant variation, suggesting that different species exhibit specific habitat preferences, rather than an even distribution across the district.

The results indicate clear habitat preferences among different bird groups. Water-associated species, such as Cattle Egrets, Kingfishers, and River Terns, were predominantly found near Kunduvada and Kondajji lakes, emphasizing the importance of wetland ecosystems in supporting avian biodiversity (Rajpar & Zakaria, 2011). Agricultural landscapes hosted species like Baya Weavers, Munias, and Indian Rollers, reflecting the role of farmlands in providing nesting and foraging grounds for these birds (Khera et al., 2009). Urban environments were mainly occupied by House Sparrows, Pigeons, and Crows, indicating that certain species are highly adaptable to human-modified habitats (Lowther, 2006). Forested and tree-dense regions, such as Devarabelakere and Hiremegalagere, supported species like Rose-Ringed Parakeets and Brahminy Starlings, suggesting that these birds prefer wooded habitats with adequate nesting sites. The presence of Indian Peafowl, a species protected under Schedule I of the Wildlife Protection Act (1972), highlights the need for continued conservation efforts in these areas.

The findings of this study highlight the need for targeted conservation strategies to protect the diverse avian populations in Davangere District. Conservation measures should include the establishment of protected bird sanctuaries, afforestation programs, and wetland conservation initiatives. Additionally, increasing public awareness and community engagement through citizen science programs can contribute to long-term monitoring and conservation efforts (Sutherland et al., 2004). Future studies should focus on seasonal bird diversity assessments, long-term monitoring of population trends, and genetic studies to understand species adaptation. Integrating remote sensing and GIS technology can further aid in mapping critical bird habitats and assessing environmental changes (Turner et al., 2003).

Table 1. List of Identified Bird Species in Davangere District

Sl. No.	Common Name	Scientific Name	Family
1	Asian Koel	*Eudynamys scolopaceus*	*Cuculidae*
2	Baya Weaver	*Ploceus philippinus*	*Ploceidae*
3	Cattle Egret	*Bubulcus ibis*	*Ardeidae*
4	House Sparrow	*Passer domesticus*	*Passeridae*
5	Indian Peafowl	*Pavo cristatus*	*Phasianidae*
6	Rose-Ringed Parakeet	*Psittacula krameri*	*Psittaculidae*
7	Brahminy Starling	*Sturnia pagodarum*	*Sturnidae*
8	Indian Roller	*Coracias benghalensis*	*Coraciidae*
9	White-Throated Kingfisher	*Halcyon smyrnensis*	*Alcedinidae*
10	Black Drongo	*Dicrurus macrocercus*	*Dicruridae*
11	Purple Sunbird	*Cinnyris asiaticus*	*Nectariniidae*
12	Common Myna	*Acridotheres tristis*	*Sturnidae*
13	Little Cormorant	*Microcarbo niger*	*Phalacrocoracidae*
14	Greater Coucal	*Centropus sinensis*	*Cuculidae*
15	Red-Wattled Lapwing	*Vanellus indicus*	*Charadriidae*
16	Eurasian Collared Dove	*Streptopelia decaocto*	*Columbidae*
17	Grey Francolin	*Ortygornis pondicerianus*	*Phasianidae*
18	Oriental Magpie Robin	*Copsychus saularis*	*Muscicapidae*
19	Spotted Dove	*Spilopelia chinensis*	*Columbidae*
20	Jungle Babbler	*Argya striata*	*Leiothrichidae*
21	Green Bee-Eater	*Merops orientalis*	*Meropidae*
22	Common Kingfisher	*Alcedo atthis*	*Alcedinidae*
23	Indian Pond Heron	*Ardeola grayii*	*Ardeidae*
24	Pied Bushchat	*Saxicola caprata*	*Muscicapidae*
25	Black Kite	*Milvus migrans*	*Accipitridae*

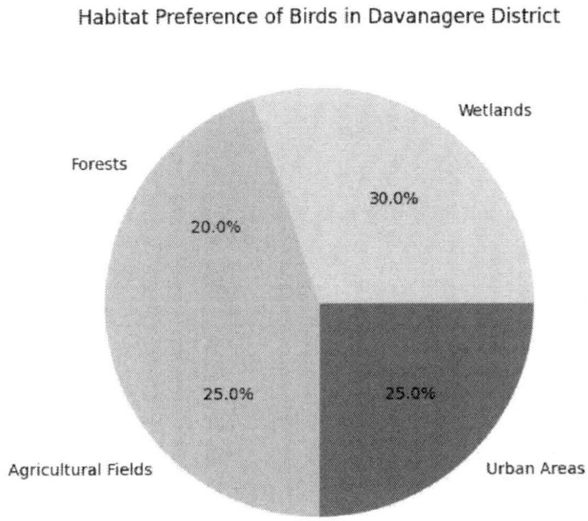

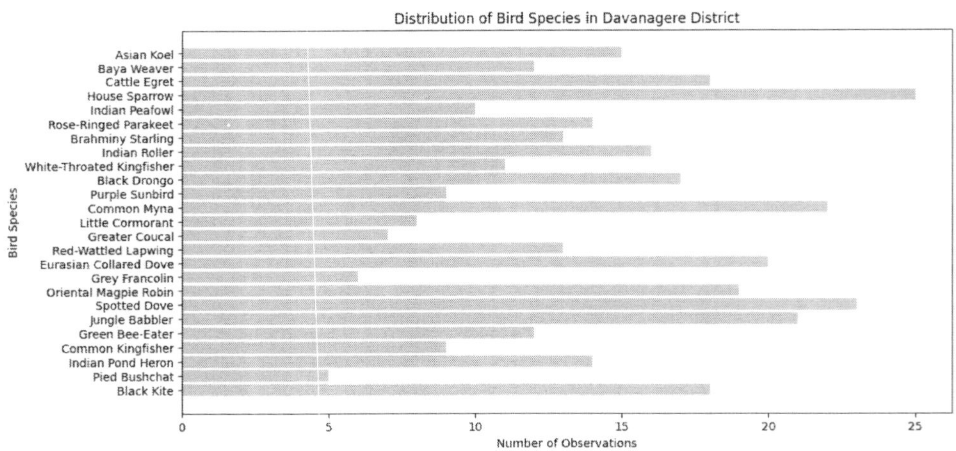

Fig: 2 The bar diagram showing distribution of bird species in Davangere

Conclusion

This study provides a comprehensive assessment of bird diversity in Davangere District, documenting 25 avian species across multiple ecological zones. The findings confirm moderate avian diversity, with clear patterns of habitat preferences among different bird groups. While the district supports a variety of bird species, ongoing threats such as habitat loss, pollution, and climate change pose significant challenges to avian conservation. To mitigate these threats, conservation efforts should focus on habitat protection, wetland restoration, and sustainable agricultural practices. Additionally, regular bird surveys and public engagement initiatives can aid in long-term biodiversity conservation. By implementing these strategies, Davangere District can continue to serve as an important habitat for a wide range of bird species, ensuring their survival and ecological contributions for future generations

References

1. Ali, S., & Ripley, S. D. (1983). Handbook of the Birds of India and Pakistan (Vols. 1–10). Oxford University Press.
2. Bibby, C. J., Burgess, N. D., Hill, D. A., & Mustoe, S. (2000). Bird Census Techniques. Academic Press.
3. BirdLife International. (2019). State of the World's Birds: Indicators for Our Changing World. BirdLife International.
4. Francis, C. D., Ortega, C. P., & Cruz, A. (2009). Noise pollution changes avian communities and species interactions. Current Biology, 19(16), 1415-1419.
5. Gaston, K. J. (2010). Valuing common species. Science, 327(5962), 154-155.
6. Khera, N., Mehta, V., & Kumar, A. (2009). Diversity and abundance of avifauna in urbanizing landscapes of India: A case study of emerging townships around Delhi. Environmental Monitoring and Assessment, 158(1), 451-461.
7. Kumar, R. S., & Jayapal, R. (2020). Checklist of Birds of Karnataka. Indian Birds, 16(5), 125-140.
8. Kumar, V., Kumar, R., & Singh, D. (2020). Impact of agricultural pollution on avian biodiversity in India. Journal of Environmental Biology, 41(2), 215-223.
9. Lowther, P. E. (2006). House Sparrow (Passer domesticus). The Birds of North America Online.
10. Magurran, A. E. (2004). Measuring Biological Diversity. Blackwell Science.
11. Mineau, P., & Whiteside, M. (2013). Pesticide acute toxicity is a better correlate of U.S. grassland bird declines than agricultural intensification. PLoS One, 8(2), e57457.
12. Mukherjee, S., Saha, G., & Ghosh, D. (2018). Conservation challenges for wetland birds: A case study from India. Journal of Ecology and Environment, 42(1), 14-21.
13. Parmesan, C., & Yohe, G. (2003). A globally coherent fingerprint of climate change impacts across natural systems. Nature, 421(6918), 37-42.
14. Praveen, J., Jayapal, R., & Pittie, A. (2016). A checklist of the birds of India. Indian Birds, 11(5&6), 113-170.

15. Rajpar, M. N., & Zakaria, M. (2011). Bird species abundance and their correlation with microclimate and habitat variables at natural wetland reserve, Peninsular Malaysia. International Journal of Zoology, 2011, 1-17.
16. Sekercioglu, C. H. (2006). Increasing awareness of avian ecological function. Trends in Ecology & Evolution, 21(8), 464-471.
17. Sutherland, W. J., Newton, I., & Green, R. E. (2004). Bird Ecology and Conservation: A Handbook of Techniques. Oxford University Press.
18. Turner, W., Spector, S., Gardiner, N., Fladeland, M., Sterling, E., & Steininger, M. (2003). Remote sensing for biodiversity science and conservation. Trends in Ecology & Evolution, 18(6), 306-314.

CHAPTER 7

ASSESSING HABITAT-BASED POPULATION DENSITY OF GREEN BEE-EATER (*MEROPS ORIENTALIS*) IN HANUMANAHALLI VILLAGE, GANGAVATHI TALUK, KARNATAKA

Krishnakumar Myala[1] and Shwetha. A.[2]

[1]*Principal and Head of Zoology, Kalmath Degree College for Science, Gangavathi – 583227.*
Koppal District. Karnataka State. Mobile No.: 9900835297.
Email Id: kkz.bio@gmail.com
[2]*Assistant Professor, Department of P.G Studies and Research in Applied Zoology, Kuvempu University, Jnana Sahyadri, Shankaraghatta, Shivamogga District, Karnataka State.*
Email Id: shweth29@gmail.com

ABSTRACT:

The green bee-eater (Merops orientalis) is a vital bioindicator species whose population distribution reflects ecological conditions. This study assesses the population density of green bee-eaters across different habitats in Hanumanahalli village, Gangavathi Taluk, Karnataka, from January 2020 to December 2022. Using the Point Count Method, observations were conducted in riverine, cropland, and urban areas.

The findings reveal significant habitat-based variations, with the highest population density in riverine zones (55.5%), followed by cropland (34.69%) and urban areas (9.81%). Statistical analysis (ANOVA, $F = 207.8$, $p = 0.0001$) confirms habitat type as a key determinant of species distribution. Factors such as food availability, nesting habitat suitability, and reduced anthropogenic disturbances significantly influence population density. Conservation measures, including habitat restoration and urban biodiversity enhancement, are recommended to support the sustainability of Merops orientalis populations in the region.

Keywords: Percentage of population, Diversity, Bird activity in different zones, Green bee-eater, Merops orientalis.

INTRODUCTION

Our planet thrives with an incredible diversity of organisms, from microscopic entities like viruses and bacteria to majestic macroorganisms such as plants and animals, all contributing to the intricate web of biodiversity (Dhindsa and Saini, 1994; Hosetti, 2008). Among these, avifaunal diversity—the variety of bird populations—plays a crucial role in maintaining ecological balance by supporting both flora and fauna. Notably, bird populations serve as key indicators of ecosystem health, as their densities often reflect the abundance of other species within the environment (Wilson and Comet, 1996; Blake, 2007; Hosetti, 2001).

The green bee-eater (Merops orientalis), a small insectivorous bird commonly found in rural India, is classified as Least Concern (LC) on the IUCN Red List (Rawlings et al., 2005). These vibrant birds are usually seen in pairs or colonies, exhibiting distinctive plumage and tail structures that blend seamlessly with surrounding vegetation (Mishra, 2019). They nest in long tunnels excavated in sandy riverbanks, with clutch sizes ranging from 3 to 6 eggs, an incubation period of approximately 14 days, and a fledging success rate of 83.7% (Asokan et al., 2010). Green bee-eaters are diurnal and highly social, spending most of their time scanning (52.5%) and feeding (21.3%), with behavioral variations observed across seasons, time blocks, and habitats (Ali and Asokan, 2015). They also roost in large groups, further emphasizing their communal nature.

Population densities of green bee-eaters vary significantly across habitats, with the highest numbers recorded along riverbanks, followed by agricultural lands and human settlements (Asokan et al., 2003). Key factors influencing their population dynamics include vegetation structure, food availability, climate conditions, and human disturbances (Asokan et al., 2003). Green bee-eaters (Merops orientalis) and their close relatives, European bee-eaters (Merops apiaster), serve essential ecological roles as insectivores and ecosystem engineers (Mishra, 2019; Gyuracz et al., 2013). Both species exhibit colony-based nesting and diverse feeding and social behaviors. While green bee-eater populations fluctuate based on local environmental conditions, European bee-eaters in Hungary maintained a stable population between 1992 and 2013, with an estimated 10,600–19,600 breeding pairs (Gyuracz et al., 2013). However, conservation challenges such as vegetation overgrowth, collapsing nesting sites, and human disturbances pose significant threats to these species (Gyuracz et al., 2013). In Uzbekistan, conservation efforts include the development of bioacoustic repellents and strategies to protect nesting colonies and prevent illegal hunting (Chorshanbiyevich et al., 2023).

This study examines the population density of green bee-eaters (Merops orientalis) in different zones of Hanumanahalli village, Gangavathi Taluk, Karnataka. As a resident bird in the study area, the green bee-eater serves as an important bioindicator, providing valuable insights into local avian ecology and environmental conditions.

Materials and methods

a. Study area

Hanumanahalli is a small village spanning approximately 5 kilometers, characterized by a diverse landscape of mountains, rivers, agricultural fields, and forests. Home to 649 residents, it falls under the administration of Sanapur Gram Panchayat in Gangavathi Taluk, Koppal District, Karnataka (Fig. 1). Situated at an average altitude of 472 meters above mean sea level (MSL), the village lies about 14 kilometers from Gangavathi Taluk and 29 kilometers from Hosapete in Vijayanagara District.

A key geographical feature of Hanumanahalli is its proximity to a large dam on the Tungabhadra River near Hosapete. The river's backwaters extend into the village, creating a thriving ecosystem with dense vegetation. This rich environment supports diverse bird species, providing ideal conditions for avian activities such as breeding and nesting (Fig. 2a, 2b, and 2c).

Karnataka State

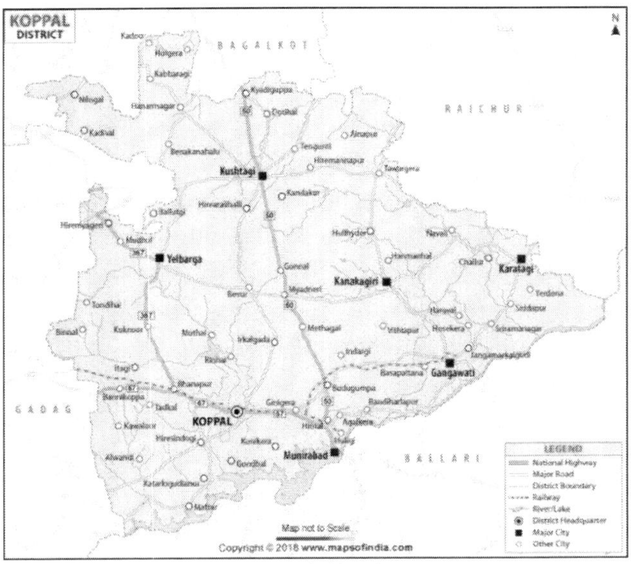

Koppal district

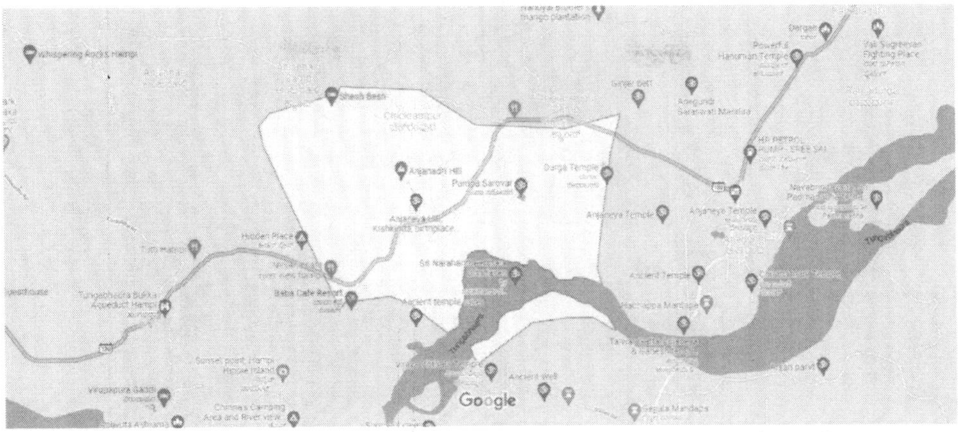

Source: ***Maps of India.***

Hanumanahalli, Gangavathi taluk

Fig: 1 Study Area Maps

2.a. Riverine area

2.b. Cropland area

2.c. Urban area

Fig: 2 Photographs of different sites

b. Methodology

For this study, data collection involved systematic weekly observations conducted over 36 months, from January 2020 to December 2022. Adopting a methodology similar to the "Avifaunal Diversity of Nagpur City, MS, India" study by Wanjari (2012), the Point Count Method was used, with surveys conducted four times a month. Observations were scheduled during two time slots - 6:30 to 10:30 AM and 4:00 to 7:00 PM to maximize bird activity detection.

Binoculars (Olympus 10 50) were used for bird identification, while reference materials, including field guides by Ali (2002) and Grimmett et al. (2011), were consulted. In cases where visual confirmation was challenging, photographs were taken to ensure precise species identification.

Multiple techniques were employed to identify birds based on their distinguishing features. Data were collected three various lands (Riverine, Cropland and Urban area). Visual counting surveys, using the Point Count Method as suggested by Crump and Scott (1994), Manley et al. (2005), Wanjari (2012) and Joshi (2014), played a key role in species documentation. Additionally, direct counting was conducted by systematically walking through the study area, following the methodology outlined by Rajashekara and Venkatesha (2010). These combined approaches facilitated a comprehensive and accurate assessment of avian diversity in the region.

For data analysis, R and SPSS statistical applications were used.

III. RESULTS

a. Observations:

In this study, we observed that the populations of Green bee-eater (*Merops orientalis*) birds vary across all different sites.

Fig: 3 Merops orientalis birds photographs collected in study area

b. Reports:

Table :I. Percentage of population density of Green bee-eater bird (*Merops orientalis*) in different zones of study area.

Sl. No.	Data collected	Urban area	Cropland area	Riverine area	Total
1	Jan. 2020- Dec. 2020	11	45	72	128
2	Jan. 2021- Dec. 2021	14	48	78	140
3	Jan. 2022- Dec. 2022	16	52	82	150
Total		**41**	**145**	**232**	**418**
% of Population Density		**9.81**	**34.69**	**55.5**	**100**

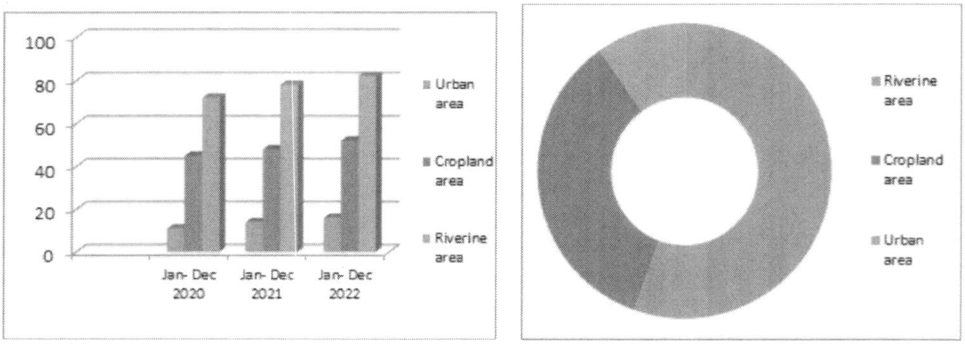

(a) (b)

Fig: 4 (a) Chart showing year wise percentage of population of Green bee-eater bird (Merops orientalis) in different zones of study area (b) Chart showing percentage of population density of Green bee-eaters from Jan. 2020 to Dec. 2022.

ANOVA analysis

The ANOVA results (Table II) reveal a significant variation in the population density of Green bee-eaters across the three sampled habitats: Riverine, Cropland and Urban areas. The F-statistic value of 207.8, along with a highly significant p-value of 0.0001, confirms that habitat type has a substantial influence on species distribution.

Among the studied habitats, riverine zones support the highest population density, indicating their ecological richness and suitability for the species. In contrast, urban areas exhibit the lowest density, likely due to habitat disturbances and environmental stressors. The strong statistical evidence underscores the necessity of conserving natural riverine ecosystems to maintain population balance and biodiversity.

Table :II. ANOVA analysis of Green bee-eater (*Merops orientalis*)

Source	Degree of freedom	Sum of Squares	Mean Square	F-stat	P-Value
Between Group	2	6096	3048	207.8	0.0001
Within Groups	6	88.00	14.67		
Total	**8**	**6184**			

Results and Discussion

The survey results indicate significant variations in the population and activity of the Green bee-eater (*Merops orientalis*) across different habitats (Table I, II, Fig. 4a and 4b). The highest population was recorded in riverine areas, with 232 birds (55.5%), followed by cropland areas with 145 birds (34.69%), while urban areas had the lowest population, with only 41 birds (9.81%) during the survey period.

Factors Influencing the Distribution of *Merops orientalis*

1. Availability of Food: Green bee-eaters are insectivorous, primarily preying on bees and other flying insects. Riverine areas provide an abundant supply of these food sources, making them highly favorable for the species (Table I and II). Similar findings have been reported by Asokan et al. (2003), whose experimental results align with these observations.
2. Nesting Habitat Suitability: The presence of loamy soil in riverine zones offers an ideal environment for nesting. Green bee-eaters prefer to burrow and nest in sandy or loamy soil, particularly along riverbanks, which serve as optimal breeding sites. This observation is consistent with the findings of Yuan et al. (2006), who documented similar nesting behavior in Merops orientalis.

Conclusion

This study underscores the vital role of habitat type in determining the population density of green bee-eaters (*Merops orientalis*), with riverine areas emerging as the most suitable environment due to their high food availability, ideal nesting conditions, and minimal anthropogenic disturbances. The significant differences observed across habitats (ANOVA, $F = 207.8$, $p = 0.0001$) highlight the species' strong preference for undisturbed ecosystems rich in insect populations. In contrast, urban areas exhibited the lowest population density, likely due to habitat fragmentation, human activities, and environmental pollution.

The findings reinforce the need for targeted conservation strategies, particularly in protecting riverine ecosystems and mitigating urbanization's impact on avian biodiversity. Initiatives such as habitat restoration, sustainable agricultural practices, and urban greening efforts can contribute to preserving *Merops orientalis* populations. Additionally, fostering community awareness and engaging local stakeholders in conservation programs will be crucial for ensuring long-term ecological balance.

As a bioindicator species, the presence and density of green bee-eaters offer valuable insights into ecosystem health, making their conservation not only essential for avian diversity but also for broader environmental stability. Future research should focus on long-term monitoring, assessing the impact of climate change, and evaluating the effectiveness of conservation measures to safeguard *Merops orientalis* and similar avian species in rapidly changing landscapes.

Acknowledgement

I sincerely express gratitude to Sri. Shivakumar Upavasi, Administrator, Kalmath Degree College for Science, Gangavathi and Sri Shama Shavi, Lecturer in Biology at Govt. P.U College, Morab, Navalagund Taluk, Dharwad District for their encouragement.

References

Ali, A.H.M.S., and S. Asokan. 2015. Diurnal-activity Patterns of the Small Bee-eater (*Merops orientalis*) in Southern India. *Tropical life sciences research,* 26(1): 09-20.

Ali, S. 2002. The Book of Indian Birds, XIII ed., Oxford University Press, Mumbai.

Asokan, S., Ali, A.M.S., and R. Manikannan. 2010. Breeding biology of the Small Bee-eater Merops orientalis (Latham, 1801) in Nagapattinam District, Tamil Nadu, India. *Journal of Threatened Taxa,* 2: 797-804.

Asokan, S., Thiyagesan, K., Negarajan, R., and R., Kanakasabai. 2003. Studies on Merops orientalis Latham 1801 with special reference to its population in Mayiladuthurai, Tamil Nadu. *Journal of environmental biology,* 24(4): 477-482.

Bennett, G.F. 1978. Avian haemoproteidae and the haemoproteids of the bee-eater family (Meropidae). *Canadian Journal of Zoology,* 56: 1721-1725.

Blake, J.G. 2007. Neotropical Forest Bird Communities: A Comparison of Species Richness and Composition at Local and Regional Scales. Condor, 109: 237-255.

Bock, C.E., and Jones, Z. F. 2004. Avian habitat evaluation: should counting birds count?. *Frontiers in Ecology and the Environment,* 2: 403-410.

Chorshanbiyevich. Y.E., Ochilovna. S.F., Rohmonkulovich. K.F., Shomurodovich. R.M., 2024. Significance and Conservation of Birds of the Genus of Bee-eater (Merops) in Uzbekistan. *Baghdad Science Journal.* 21(2): 305-312.

Crump, M.L. and Scott, N.J. 1994. Visual encounter surveys. In: Measuring and Monitoring Biological Diversity: Standard Methods for Amphibians (eds., Heyer, W. R., Donnelly, M.A., Mc Diarmid, R. W., Hayek, L. C. and M. S. Foster), Washington: Smithsonian Institution Press, 84-92.

Dhindsa, M.S. and Saini, H.K. 1994. Agricultural ornithology: An Indian perspective, Journal of Biosciences, 19: 391-402.

Grimmett, R., Inskipp, C. and Inskipp, T. 2011. Birds of the Indian Subcontinent. II ed., Oxford University Press, India.

Gyuracz. J., Nagy. K., Fuisz. T.I., Karcza. Z., and Szep. T., 2013. European Bee-eater (*Merops apiaster* Linnaeus, 1758) in Hungary: a review. *Ornis Hungarica.* 21(2):1-22

Hosetti, B.B. 2001. Glimpses of Biodiversity, Daya Publishing House, Delhi.

Hosetti, B.B. 2008. Concepts in Wildlife Managements, III ed., Daya Publishing House, Delhi.

Joshi, P.S. 2014. Diversity and Population Dynamics of Ophidian Fauna from Buldhana District, Maharashtra (India). Ph.D., Thesis, SGBAU, Amravati.

Manley, P.N., Horne, B.V., Roth, J.K., Zielinski, W.J., McKenzie, M.M., Weller, T.J., Weckerly, F.W. and Vojta, C. 2005. Multiple species inventory and monitoring technical guide. Version 1.0 (Pre-print), *USDA Forest Service,* 1-193.

Mishra, S. 2019. The Ecology of Merops Orientalis. *International Journal of Scientific Research in Biological Sciences,* 13(2).

Rajashekara, S. and Venkatesha, M.G. 2010. The diversity and abundance of water birds in lakes of Bangalore city, Karnataka, India. *Biosystematica*, 4(2): 63-73.

Rawlings. N.D., Morton. F.R., and Barrett.A.J., 2005. Merops: the peptidase database, *Nucleic Acids Research.* 34: 270-272.

Wanjari, P.D. 2012. Avifaunal diversity of Nagpur City, MS, India. *Bionano Frontier,* 5(2-I): 124-126.

Wilson, M.F. and Comet, T.A. 1996. Bird communities of northern forests: Patterns of diversity and abundance. *The Condor,* 98(2): 337-349.

Yuan, H. W., Brent Burt, D., Wang, L. P., Chang, W. L., Wang, M. K., Chiou, C.R. and Ding, T. S. 2006. Colony site choice of blue tailed bee-eaters: Influences of soil, vegetation, and water quality. *Journal of Natural History, 40*(7-8), 485-493.

CHAPTER 8

FRUGIVOROUS BAT PTEROPUS GIGENTEUS ROOST IN DAVANGERE DISTRICT, KARNATAKA: A CASE FOR PRESERVATION AS A HERITAGE SITE®

Harish Kumar T. S.[1*] and Parimala B.[2]

[1]Department of Zoology, D.R.M. Science College, Davangere, Karnataka, India-577 004.*
[2]Department of Zoology, University College of Science, Tumkur University, Tumakuru, Karnataka, India-572103.
Corresponding Email: harishsrys@gmail.com

ABSTRACT:

Bats are the only mammals that are capable of flying. They belonging to the Class Mammalia; Order: Chiroptera. In the present study, Pteropus species was reported in two different sites Davanagere district. More number of individuals of bats were recorded in the urban than in rural area and also, they prefer the Ficus tree than any other trees present in the vicinity. The behavior like feeding, sleeping, self-grooming and mating were studied.

Keywords: *Pteropus*, Honnali, Davangere, Roosting, *Ficus*, conservation.

Introduction

Bats belonging to the Order: Chiroptera and Class: Mammalia. They are the only species of mammals that are capable of flying continuously. They are distributed in the different parts of the world, except regions characterized by extremely cold climates, this phenomenon is observed globally. Bats are the second largest order of mammals (Voigt and Kingston, 2016). Bats provide highly important ecological services such as insect-pest predation, pollination and seed dispersal, thus providing valuable ecosystem services. Bats are the animals which are the least concerned group. There are many species of bats are reported from different parts of the world including India. Mainly, there are two groups of bats i.e. Megachiropterans and Microchiropterans. Megachiropterans are large in size, feed on fruits and seeds. They are commonly called frugivorous bats. Microchiropterans are smaller size, feed on various kinds of insects. They are commonly called as insectivorous bats (Boro, 2018). There are about 200 species of bats were recorded in more than 60 different countries around the world. Of these, 25% are frugivorous bats and 88% are exclusively tropical (Mistry, 2000).

The preferable habitat of frugivorous bats includes fruit canopy, open clearings and forest tracts. Most of the species are listed as Vulnerable, Endangered or Critically Endangered by the International Union for the Conservation of Naturem, which play an important role in the ecosystem functions of tropical forests, through pollination and seed dispersion (Hengjan, 2018). They are a key group in seed dispersal (Giannini and Kalko,2004), in pollination and predation, and controlling a wide variety of prey (Kunz and Fenton, 2003). The major threats for bats are loss of habitat, climate change, invasive species and illegal wildlife trade. In recent years, researchers have revealed that, bats play an important role in many ecosystems as insect predators, seed dispersers and pollinators. Today, bats are under various threat due to widespread habitat destruction, climate change, invasive species and other stresses. Without concerted international action, their populations will continue to fall, driving many species to extinction (Srinivasalu and Srinivasulu, 2001). Assessing diversity and distribution of bats in relation to land-use and anthropogenic threats in the southern Western Ghats was recorded by Deshpande (2012). There exists a huge gap relating to the research studies on bats of Davangere District. This study focused on documentation on the colony size. As most of the bat species are highly mobile and colony may base on maternity, transient, swarming and hibernating periods (Kunz et al., 2009). This study provides baseline information about the roosting of this species in Davangere District which will be beneficial for its future conservation management.

Materials and Methods

a. Study area

The study was conducted in the places like Honnali (14 .25'N; 75 .67'E) and

Davangere (14 28' N; 75 59'E) (Fig. 3 and 4), Karnataka, India. The average elevation is 540 m. The climatic condition of the district is moderate to humid monsoon type. The southwest monsoon season starts from June to September. The temperature ranged between 26 to 28 C.

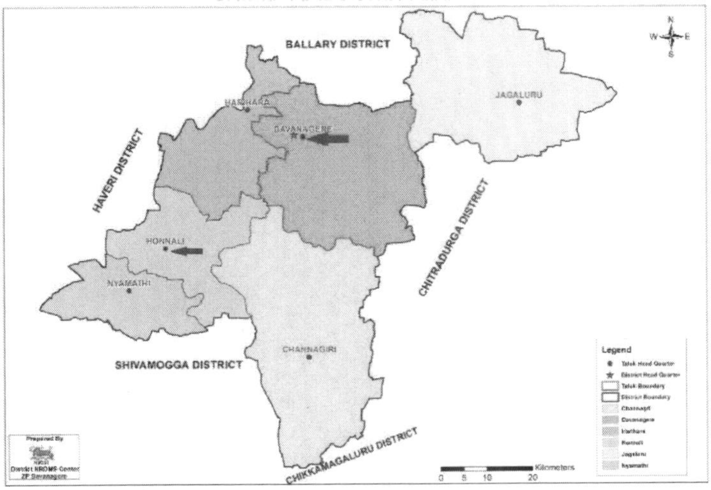

Fig: 1 Location of study area

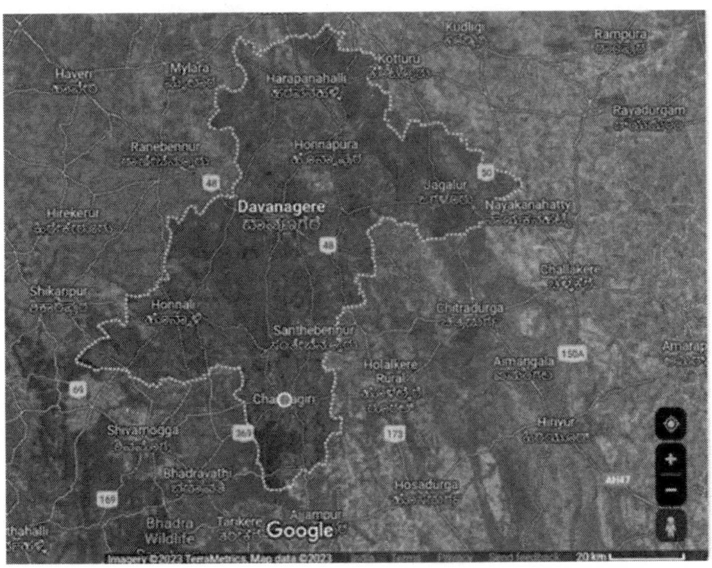

Fig: 2 Google map showing the study area

Data Collection

The population of bats were estimated by counting the number of bats visiting the habitats during dusk (1800 to 1900hr), night (2200 to 2300hr) and dawn (0500 to 0600 hr) using binocular. The bats were counted at weekly intervals from August 2022 to September 2023. Flying foxes are known to have both single roosts where large congregations occur on a single tree and scattered roosts where the individuals roost on various trees in a specific area. The behaviour like feeding, sleeping, self-grooming and mating were studied.

Fig: 3 Study area: Honnali

Fig: 4 Study area: Honnali

Fig: 5 *Pteropus gigenteus* in the branches of *Ficus* tree

Results and Discussion

The study revealed, the occurrence of frugivorous bat *Pteropus giganteus,* commonly called as Indian flying fox. *Pteropus* is largest bat species in India. There are about 1000 to 2000 individuals were hanging down in the branches of Ficus trees in the study area. These are available throughout the year (Pre-monsoon, monsoon and post-monsoon) with constant number. Ficus trees are large with greater height, diameter, crown length and crown spread with an age of more than 100-120 years old. During day time, bats are roosting in Ficus tree and in the late evening they are migrating in search of food. In the early morning, they will be back to their roosting site. Even though, there are, *Caesalpinia pulcherima, Azadirachta indica* and other trees present in the vicinity (**Fig. 3 and 4**). They use to move along the tree branches without flying, sometimes they would fly around the roosting site itself. This behaviour is more common in the early morning after arrival to the roosting site. The movement of *Pteropus* bats is greatly dependent on type of food plants. Since they do not hibernate, they consume food throughout year (Hengjan, 2018).

During mating, both male and female produce some kind of noise. Males may mark a tree with chin and neck glands during mating season (Kulzer 1990). Males in managed care make perches or walls with nick and chin glands (Kadambari, 2012). In males of many *Pteropus* species, glands on the back of the neck excrete sticky substance that mats the hair in mating season. Bats produce strong vocalization in the roosting sites during day time especially in the early morning. Bats frequently produce the vocalization when it is aggressive and fighting between two individuals (Kulzer 1990).

In both the sites, the bat population was noticeable. More number of bat species were observed in Davangere than in Honnali as the number of roosting trees were more and also the large sized old trees in Davanagere site. The trees which are present in the urban ecosystem provide a variety of services including biodiversity conservation, removal of atmospheric pollutants, oxygen generation, noise reduction, microclimate regulation, soil erosion and carbon sequestration (Srinivasulu and Srinivasulu, 2001). Trees also have the potential to make a marked improvement on air quality by absorbing carbon dioxide and other pollutants.

Table 1: Tree species used by *Pteropus giganteus* for roosting in Honnali.

Sl No.	Tree species	Common Name	No. of Trees with Bat Roosts	No of Bats (Mean ± SD)
1	*Ficus*	Fig	1	218.15±28.12
2	*Ceasalpinia*	Peacock flower tree	2	550.45±22.09
3	*Azadirachta*	Neem	1	150.10±12.00
4	*Pongamia*	Hongay (Honge)	1	100.2±45.11
5	*Polyalthia*	Ashok	2	50.76±19.0

Table 2: Tree species used by *Pteropus giganteus* for roosting in Davangere

Sl No.	Tree species	Common Name	No. of Trees with Bat Roosts	No of Bats (Mean ± SD)
1	*Neolamarckia*		1	2000.12±12.23
1	*Ficus*	Fig	1	218 .19±3.65
2	*Ceasalpinia*	Peacock flower tree	15	2500.1±2.99
3	*Azadirachta*	Neem	3	75.87±32.1
4	*Artocarpus*	Jackfruit	2	14.54±12.34
5	*Polyalthia*	Ashok	10	87.00±23

References

1. Boro, A.R., P.K. Saikia and Saikia, U. (2018). New records of bats (Mammalia: Chiroptera) from Assam, northeastern India with a distribution list of bat fauna of the state. *Journal of Threatened Taxa* 10(5): 11606–11612; http://doi.org/10.11609/jott.3871.10.5.11606-11612.
2. Giannini, N. P. & Kalko, E. K. V. (2004). Trophic structure in a large assemblage of phyllostomid bats in Panama. Oikos 105(2):209-220.
3. Hengjan Y., Saputra V., Mirsageri M., Pramono D., Kasmono S., Basri C., Ando T., Ohmori Y., Agungpriyono S. and E. Hondo (2018). Nighttime behavioural study of flying foxes on the southern coast of West Java, Indonesia. *J Vet Med Sci.* 80(7): 1146–1152.
4. Kadambari D. (2012) Assessing diversity and distribution of bats in relation to land-use and anthropogenic threats in the southern Western Ghats, India Project ID: 9686-1, The Ruffor small Grant Foundation.
5. Kunz, T. H. & Fenton, M. B. (2003). Bat Ecology. Chicago, The University of Chicago Press, Chicago.778p.
6. Kunz, T.H., M. Betke, N.I. Hristov, and M. Vonhof (2009). Methods for assessing colony size, population size, and relative abundance of bats. In: Ecological and behavioural methods for the study of bats, 2nd ed, eds. T.H. Kunz and S. Parsons, 133–157. Baltimore: Johns Hopkins University Press., 2012
7. Srinivasulu C. and B. Srinivasulu (2001). Bats of the subcontinent-Anupdate. Current Science, 80:1378-1380.
8. Mistry S. (2000). Tropical Ecosystems: Structure, Diversity and human welfare. In: Ganeshiah K. S., R. Umashankarand K. S. Bawa (Eds.) Proceedings of the International Conference on Tropical Ecosystems, Oxford-IBH, New Delhi, pp 707-710.
9. Voigt, C. C. & Kingston, T. (2016). Bats in the Anthropocene. In: Voigt, C. C. & Kingston, T. eds. Bats in the Anthropocene: Conservation of Bats in a Changing World. New York, Springer International Publishing, p. 1-9.

CHAPTER 9

STUDY ON PLANT REMEDIES FOR DIABETES

Shivanand Payamalle

Department of Botany, KLE Society's
Gudleppa Hallikeri Arts, Science and Commerce College
Haveri, Karnataka-581110
Email id: shivanand.payamalle0@gmail.com
Mob. No: +919591703071

ABSTRACT:

Diabetes is the abnormality in the production of insulin ant its action on the regulation of glucose level in the body. Diabetes may lead to the serious problems related to the health such as heart disease, stroke, kidney failure, blindness chronic infection etc. it is one of the predominant disease of the world affect on majority of the population. To minimize the effect of the diabetes plant based medicines are used worldwide, may be entire plants or used in the form of decoction or infusion. The plants show the antidiabetic activity due to the presence of phytochemicals such as- flavanoids, phenols, alkaloids, terpenoids, and glucosinolates. These chemicals play an important role in controlling the diabetes along with them also possess antioxidant, anti-inflammatory, and antimicrobial activities. Traditional system of medicine is the method used in the treatment of various diseases including diabetes, this system of medicine is practiced throughout the world since from longer period. This report is highlighting the uses of the plants against the diabetes and action of phytochemicals to reduce the risk of diabetes. Further concern and more research is required to minimize the effect of diabetes on the health and lead healthy lifestyle.

Keywords: Diabetes, traditional medicine, Phytochemicals, antidiabetic.

Introduction

Diabetes mellitus is a long-standing endocrine disorder that negatively changes the metabolism of carbohydrates, proteins, fat electrolytes and water, largely because of an abnormality in insulin production (1). It is linked with long term complications like heart disease, stroke, kidney failure, blindness, nerve damage and neuropathy, atherosclerosis, chronic infection, immune deficiency and peripheral vascular disease which can result in ulcers, gangrene and amputation (2, 3). There have been reports of two main forms of diabetes mellitus, i.e., Type 1 and Type 2 (4, 5). The type 1 diabetes is largely caused by inadequate secretion of insulin leading to sustained hyperglycemia (4). Type 2 diabetes mellitus is shown to be higher than all other forms of diabetes and occurs in around 90% of the world's population (6), caused by ineffective metabolic functions or damage or malfunction to the organs. Bahrami and Gerstein (7) added that type 2 diabetes mellitus is a severe public health mense because 50% of individuals suffering from it remain undiagnosed. An estimate of approximately 80 % of such individuals reside in developing countries (8, 9). Blood glucose management is key in the control of the condition and the avoidance of related complications.

Although traditional therapies, such as insulin therapy and oral hypoglycemics, are important, interest has been growing in the prospects of phytochemicals—bioactive plant compounds—for the control of diabetes. Phytochemicals are plant compounds that are responsible for the color, flavor, and resistance to diseases in plants. These compounds are usually grouped into a number of categories, such as alkaloids, flavonoids, phenolic acids, terpenoids, and glucosinolates. Most of these phytochemicals have been found to possess antioxidant, anti-inflammatory, and antimicrobial activities, the majority of which are pertinent in the control of diabetes.

Phytochemicls and diabetes

Flavonoids

Flavonoids are polyphenolic compounds that are commonly present in fruits, vegetables, tea, and herbs. Some flavonoids, such as quercetin, kaempferol, and catechins, have been studied for their potential in controlling blood sugar levels. These substances have antioxidant activity, increase insulin sensitivity, and modulate glucose metabolism. Flavonoids lower oxidative stress by neutralizing free radicals, which is important because oxidative stress is a factor in insulin resistance in diabetic patients.

Certain flavonoids, including quercetin, aid in enhancing insulin sensitivity by activating AMP-activated protein kinase (AMPK), which is an essential mediator of glucose and lipid metabolism (10). A Supplementation with quercetin in diabetic rats enhanced blood glucose control and increased insulin sensitivity (11). Further studies by Dey et al. (12) indicated that catechins present in green tea can decrease blood glucose levels and alleviate insulin resistance.

Phenolic Acids

Phenolic acids, such as chlorogenic acid and caffeic acid, are found in coffee, fruits, and vegetables. These substances have high antioxidant activity and can contribute to the regulation of blood sugar. Phenolic acids can suppress enzymes like alpha-glucosidase, which are involved in the digestion of carbohydrates into glucose, thus slowing the increase in blood sugar after meals. Phenolic acids have been reported to enhance insulin sensitivity by lowering inflammation and oxidative stress in the body.

A clinical trial showed that chlorogenic acid, which is present in coffee, lowered postprandial blood glucose levels in type 2 diabetic patients by a significant amount (13). Ong et al., (14) stated that chlorogenic acid enhanced insulin sensitivity in obese individuals.

Alkaloids

Alkaloids are plant compounds with nitrogen, which include bitter melon (Momordica charantia), berberine (genus Berberis), and ginseng. These have been reported to have antidiabetic effects. Certain alkaloids, such as those present in bitter melon, can act like insulin by stimulating the uptake of glucose into cells and increasing insulin release from the pancreas.

Berberine has been found to control blood glucose by enhancing insulin sensitivity, lowering hepatic glucose output, and enhancing skeletal muscle glucose uptake. Berberine, a strong alkaloid, significantly lowered blood glucose levels and enhanced lipid profiles in type 2 diabetic patients (15). Another research by Wang et al. (16) identified the antidiabetic action of bitter melon in experimental animals, paving the way for its use in diabetic treatment.

Terpenoids

Terpenoids are an array of chemical compounds in herbs such as ginseng, turmeric, and ginger. They have emerged as potential molecules to lower blood glucose levels and enhance insulin sensitivity. Terpenoids, including curcumin in turmeric, exert strong anti-inflammatory and antioxidant activities that can suppress the chronic inflammation seen with insulin resistance.

Certain terpenoids, such as ginsenosides in ginseng, stimulate major enzymes responsible for glucose metabolism, such as AMPK, which improves insulin sensitivity and glucose uptake. Earlier study found that ginsenosides in ginseng had a profound effect on insulin resistance and lowering blood glucose levels in type 2 diabetes patients (17). Seo et al. (18) also showed that curcumin, a terpenoid, enhanced insulin sensitivity and lowered blood glucose levels in both animal and human models.

Glucosinolates

Glucosinolates are sulfur-containing phytochemicals in cruciferous vegetables such as broccoli, cauliflower, and kale. Glucosinolates are believed to exert antidiabetic action primarily through their antioxidant and anti-inflammatory activities. Glucosinolates inhibit inflammation by regulating inflammatory cytokines, which may prevent insulin resistance.

The breakdown products of glucosinolates, including sulforaphane, have been reported to enhance insulin sensitivity and suppress oxidative stress. Truong et al. (19) identified that sulforaphane, which is a product of glucosinolates, enhanced insulin sensitivity and alleviated oxidative stress in diabetic mice. Chen et al. (20) further demonstrated the efficacy of cruciferous vegetables in controlling diabetes based on their glucosinolate composition.

Traditional system to heal the diabetes

The earlier workers stated that medicinal plants are employed in the treatment of diabetes primarily because they contain active compounds with potent activities against the disease (21, 22). The findings obtained by the Rajesham et al. (23) indicated that medicinal plants that are employed against the treatment of diabetes not only yielded good result but also exhibited effective activity compared to synthetic medicine.

In recent years there has been growth at an exponential rate in the area of herbal medicine and such drugs are increasing in popularity in both developing and developed nations due to their natural origin and reduced side effects (24). Most traditional medicines being used today are extracted from medicinal plants, and organic materials (25).

Several medicinal plants, which are used for more than 1000 years as rasayana are found in herbal formulations of Indian traditional systems of health care (26). Most practitioners of indian systems of medicine prepare and distribute their own species (27). World Health Organization (WHO, 2024) has enlisted 21,000 plants, which are employed for the purpose of medicines globally. Out of these 2500 species, there are in India, and 150 species are commercially utilized on a quite large scale (24). Earlier research work on the indigenous medicine on diabetes indicated that there are approximately 29 plant species from 23 families that have been utilized for diabetes in decoction form (28). Agyei et al (29) indicated the application of 39 plants against the hypertension and diabetes. Herbal preparation derived from the Moringa stenopetala (33.5 %), *Trigonella foenumgraecum* (27.4 %), and Thymus schimperi and found to be effective on the researched patients (30).

Deshmukh et al (31) indicated ten medicinal plants with antidiebetic activity because of the availability of phytochemicals within them. In India too various plants are utilized against the diabetes by preparation of drug with definite composition of plant products, over 600 plant species are utilizing throughout the nation against the diabetes (32, 24). Asian food system and medicinal system have shown clinical and experimental antidiabetic efficacy focused on Type 2 diabetes (32). The decoction, infusion, maceration techniques, food or raw plant cooking are employed against the diabetes are proved to be effective (34). The medicinal plants are proven to be good remedies for the treatment of various diseases including diabetes with less side effects and best uses.

Conclusion

Diabetes is one of the serious health issues affecting to the most of the people. It is causing some of the side effects such as kidney failure, blindness; heart disease, chronic illness, etc. For the treatment of diabetes various plant based remedies have showed significant effect in the managing glucose level in diabetes patients. There are several plants which used for the curing of diabetes in the form of raw plant, decoction or infusion, etc. These plants regulate blood sugar level, improve sensitivity of insulin, and reduce complications related to diabetes. These plants show effective action against due to the presence of various phytochemicals such as- alkaloids, phenols, flavanoids, terpenoids etc. each active compound shows different ways of mode of action in the management of glucose level. Although plants have showed their activity either in laboratory or by the traditional healers, but more concern and research to be done to uncover the active ingredient present in plants and proper medication to be produced so that the diabetes can be treated at grass root level.

References

1. F. Farzaei, M.R. Morovati, F. Farjadmand, M.H. Farzaei, A mechanistic review on medicinal plants used for diabetes mellitus in traditional Persian medicine, Journal of evidence-based complementary & alternative medicine 22 (4) (2017) 944–955.
2. J. W. J. Beulens, F. Rutters, L. Ryd´en, O. Schnell, L. Mellbin, H. E. Hart, R. C. Vos, Risk and management of pre-diabetes. European Journal of Preventive Cardiology 26 (2_Supplement) (2019) 47–54.
3. J. Boulton, L. Vileikyte, l. G. Ragnarson-Tennval, J. Apelqvist, The global burden of diabetic foot disease. Lancet 366 (9498) (2005) 1719–1724.
4. I. Chukwuma, R. Mopuri, S. Nagiah, A. A. Chuturgoon, M. S. Islam, Erythritol reduces small intestinal glucose absorption, increases muscle glucose uptake, improves glucose metabolic enzymes activities and

increases expression of Glut-4 and IRS-1 in type 2 diabetic rats. Eur. J. Nutr. (2017), https://doi.org/ 10.1007/s00394-017-1516-x.

5. Z. Hu, F. Gao, L. Qin, Y. Yang, H. Xu, A case-control study on risk factors and their Interactions with prediabetes among the elderly in rural communities of yiyang city, hunan province. J. Diabetes Res. (2019) 1–8, https://doi.org/10.1155/2019/1386048.
6. R.E. Pratley, The early treatment of type 2 diabetes. Am. J. Med. 126 (2013) S2–S9.
7. J. Bahrami, H. Gerstein, The prevalence of undiagnosed type 2 diabetes on the general medicine ward. Can. J. Diabetes 40 (5) (2016) S49–S50, https:// doi. org/10.1016/j.jcjd.2016.08.139.
8. H. Jangid, S. Chaturvedi, M. Khinchi, An Overview on Diabetes Mellitus. Asian Journal of Pharmaceutical Research and Development. 5(3) (2017) 1-11.
9. World Health Organisation (WHO), Diabetes, 2024. https://www.who.int/news-room/fact-sheets/detail/diabetes (Accessed 5 January 2025).
10. Y. Zhang, Q. Wu, J. Liu, Z. Zhang, X. Ma, Y. Zhang. J. Zhu, R. W. Thring, M. Wu, Y. Gao, H. Tong, Sulforaphane alleviates high fat diet-induced insulin resistance via AMPK/ Nrf2/GPx4 axis. Biomedicine & Pharmacotherapy 152 (2022) 113273. https://doi.org/10.1016/j.biopha.2022.113273.
11. R. Dhanya. Quercetin for managing type 2 diabetes and its complications, an insight into multitarget therapy. Biomedicine & Pharmacotherapy 146 (2022) 112560. Contents lists available at ScienceDirect Biomedicine & Pharmacotherapy journal homepage: www.elsevier.com/locate/biopha https://doi.org/10.1016/j.biopha.2021.112560
12. *P. Dey, G. Y. Sasaki, P. Wei, J. Li, L. Wang, J. Zhu, D. McTigue, Z. Yu, R. S. Bruno, Green tea extract prevents obesity in male mice by alleviating gut dysbiosis in association with improved intestinal barrier function that limits endotoxin translocation and adipose inflammation. J. Nutr. Biochem. 67 (2019) 78–89. doi: 10.1016/j.jnutbio.2019.01.017.*
13. Y. Yan, Q. Li, .L. Shen, K. Guo, X. Zhou. Chlorogenic acid improves glucose tolerance, lipid metabolism, inflammation and microbiota composition in diabetic db/db mice. Front Endocrinol (Lausanne). (2022) 17;13:1042044. doi:10.3389/fendo.2022.1042044.
14. H. Dong, N. Wang, L. Zhao, F. Lu, "Berberine in the treatment of type 2 diabetes mellitus: A systematic review and meta-analysis." Evidence-Based Complementary and Alternative Medicine. (2012), 591654. doi:10.1155/2012/591654

15. *K. W. Ong, A. Hsu, B. K. Tan. Anti-diabetic and anti-lipidemic effects of chlorogenic acid are mediated by AMPK activation. Biochemical Pharmacology. 85(9) (2013)1341–1351. doi: 10.1016/j.bcp.2013.02.008.*

16. H. Y. Wang, W. C. Kan, T. J. Cheng, S. H. Yu, L. H. Chang, J. J. Chuu. Differential anti-diabetic effects and mechanism of action of charantin-rich extract of Taiwanese *Momordica charantia* between type 1 and type 2 diabetic mice. Food and Chemical Toxicology. 69 (2014), 347-356. https://doi.org/10.1016/j.fct.2014.04.008

17. P. Zhou, W. Wie, S. He, Y. Sun, X. Meng, G. Sun, X. Sun. Ginsenoside Rb1 as an Anti-Diabetic Agent and Its Underlying Mechanism Analysis Cells. 8(3) (2019)204. doi: 10.3390/cells8030204

18. K. Seo , M. S. Choi, U. J. Jung, H. J. Kim, J. Yeo, S. M. Jeon M. K. Lee . Effect of curcumin supplementation on blood glucose, plasma insulin, and glucose homeostasis related enzyme activities in diabetic db/db mice. Mol. Nutr. Food Res. 2008, 1 52, 000 – 000 DOI 10.1002/mnfr.200700184

19. T. T. P. Truong, T. Koyama, Glucosinolate Sinigrin Improves Insulin Resistance to Suppress Glutathione Consumption in Type 2 Diabetic Mice. J Diabetes Metab. 11:859 (2020). doi: 10.35248/2155-6156.20.11.859.

20. G. C. Chen, W. P. Koh, J. M. Yuan, L. Q. Qin, R.M. van Dam. Green Leafy and Cruciferous Vegetable Consumption and Risk of Type 2 Diabetes: Results from the Singapore Chinese Health Study and Meta-Analysis. *Br. J. Nutr.* 119 (9) (2018), 1057–1067. doi:10.1017/s0007114518000119.

21. W. Kooti, M. Farokhipour, Z. Asadzadeh, D. A. Larky, M. A. Samani. The role of medicinal plants in the treatment of diabetes: a systematic review. Electronic Physician. 8(1) (2016) 1832-1842, DOI: http://dx.doi.org/10.19082/1832.

22. B. Jacob, R. T. Narendhirakannan. Role of medicinal plants in the management of diabetes mellitus:a review. 3 Biotech (2019) 9:4 https://doi.org/10.1007/s13205-018-1528-0.

23. V. V. Rajesham, A. M. Aparna, P. Kalyani, M. Balaji, K. S. Reddy, T. Rama Rao, "Exploring Herbal Remedies for Diabetes Management: A Comprehensive Review of Efficacy, Epidemiology, and Quality of Life Considerations," International Research Journal of Pharmacy and Medical Sciences, 7 (2) (2024). 67-73.

24. M. Modak, P. Dixit, J. Londhe, Saroj Ghaskadbi, T. P. A. Devasagayam. Indian Herbs and Herbal Drugs Used for the Treatment of Diabetes. J. Clin. Biochem. Nutr. 40 (2) (2007) 163-173

25. J. K. Grover, S. Yadav, V, Vats. Medicinal plants of India with antidiabetic potential. J. Ethnopharmacol. 81 (2002) 81-100.

26. P. Scartezzini, E. Sproni. Review on some plants of Indiam traditional medicine with antioxidant activity. J. Ethnopharmacol. 71 (2000) 23-43.

27. S. D. Seth, B. Sharma. Medicinal plants of India. Indian J. Med. Res. 120 (2004) 9-11.

28. I. Hinad, Y. S'hih, A. Elhessni, A. Mesfioui, M. l. Ouahidi. Medicinal plants used in the traditional treatment of diabetes in Ksar Elkebir Region (North-Western Morocco). Pan African Medical Journal. 42(319) (2022). 10.11604/pamj.2022.42.319.32572

29. T. A. Agyei, Y. Appau, K. B. Barimah, A. Asase. Medicinal plants used for management of diabetes and hypertension in Ghana. Heliyon 9 (2023) e22977

30. G. D. Lema, E. D. Gebeyaw, Z. A.Yferu, S. F. Mulatu, A. B. Dagnaw, G. B. Aydagnuhm, E. A. Ayicheh. Herbal medicine use and its impact on glycemic control among diabetes patients at governmental hospitals in Debre Berhan, Ethiopia: A cross-sectional study. Metabolism Open 23 (2024) 100311.

31. P. Deshmukh, A. Phuse, S. Nimkar, D. Ruikar, A. Raut. A review: Anti-Diabetic Plants and its Recent Application. Journal of Ayurvedic and Herbal Medicine. 10(2) (2024)56-66.

32. N. Goel, U. Tiwari, S. Singh, M. Gupta, S. P. Gupta, Indian Medicinal Plants Utilized in Traditional Medicine for the Management of Diabetes. IJBPAS, 13(10) (2024): 5209-5225.

33. P. K. Perera, Y. Li. Functional herbal food ingredients used in type 2 diabetes mellitus. Pharmacognosy Reviews 6 (11) (2012). DOI:10.4103/0973-7847.95863.

34. M. Belmouhoub, M. Tacherfiout, F. Boukhalfa, Y. K. Khodja, M. B. Bey, Traditional medicinal plants used in thetreatment of diabetes: Ethnobotanical and ethnopharmacological studies and mechanisms of action. Int. J. Plant. Bas. Pharm, 2(2) (2022), 145-154.

CHAPTER 10

INVESTIGATION OF THE DRAWIDA MODESTA LIFE CYCLE WITHIN CONTROLLED LABORATORY ENVIRONMENTS

Reshma Banu

Department of Zoology, Shri Shankar Anand Singh Government First Grade College, Hosapete - 583201
Email: reshmak1221@gmail.com

ABSTRACT:

This study investigated the life cycle of Drawida modesta, a topsoil endogeic earthworm species, over a one-year laboratory rearing period. D. modesta exhibits low biomass (3.48 mg), inhabits pasture ecosystems, and is distributed within the upper 10-15 cm of the soil profile. Morphological features include light dorsal pigmentation and a body length of 42-50 mm. The species is Phytophagous and demonstrates a semi-continuous cocoon production cycle, with clitellum development occurring within 27-32 days. Reproduction is via cross-fertilization, yielding an average of 27 cocoons per individual annually, with each cocoon producing 2-3 hatchlings after a 26-day incubation period. Seasonal variations in cocoon production were observed, with peak activity during summer, moderate production during monsoon, and minimal production during winter. These findings contribute to a comprehensive understanding of D. modesta reproductive strategies and its potential applicability in vermiculture.

Keywords: *Cocoons, Earthworm , Drawida modesta, Hatchlings, Lifecycle.*

Introduction

Earthworms, belonging to the kingdom Animalia, phylum Annelida, class Chaetopoda, and order Oligochaeta, are multicellular, eucoelomate invertebrates that inhabit terrestrial environments. Originating approximately 650-570 million years ago during the Precambrian era (Bouche, 1983; Piearce et al., 1990), they constitute a significant component of soil macrofauna, often representing up to 80% of below-ground macro invertebrate biomass (Lavelle et al., 1995; Turbe *et al.*, 2010). While ubiquitous, their distribution is limited in certain habitats, such as cold regions, rocky terrains, and areas with sparse soil and vegetation (Edwards and Bohlen, 1996). Species with widespread distribution are termed peregrine, whereas those confined to specific environments are considered endemic (Edwards and Lofty, 1972). Earthworm diversity, abundance, and distribution patterns serve as indicators of soil characteristics and climatic conditions within a given geographical region. Historically recognized as "unheralded soldiers of the soil" (Darwin) and the "intestine of the Earth," earthworms dominate soil invertebrate communities in terms of biomass and overall activity, particularly in temperate and tropical soils.

Oligochaetes are hermaphroditic, with spermatogenesis occurring in the testes, followed by sperm storage in the seminal vesicles and testes sacs. Oocytes are released from the ovaries and stored in the ovisacs. Despite hermaphroditism, earthworms typically engage in cross-fertilization, although parthenogenetic reproduction has been observed in some species. The reproductive strategy of earthworms is notable due to their hermaphroditic nature (Kale et al., 1982; Julka, 1988; Kaushal and Bisht, 1992; Kaushal et al., 1995).

The reproductive biology and life cycle of earthworms are critical areas of study in both ecology and agriculture. Recent research has illuminated key aspects of their reproduction, including morphology, mating, fertilization, cocoon production, and hatching. Foundational work by Bahl (1922, 1927) on Indian *Pheretima* species established the importance of seasonal influences on reproductive activity. Subsequent studies by Stephenson (1930) and others on Lumbricidae further developed our understanding of their life cycles and ecological roles.

Investigations into the environmental requirements of composting earthworms, such as temperature and moisture, have been crucial (Kaplan et al., 1980; Reinecke and Krief, 1981; Viljoen and Reinecke, 1989). Research on *Eudrilus eugeniae* has provided detailed insights into population dynamics and fecundity (Kale and Bano, 1991, 1994).

The influence of diverse organic substrates on earthworm growth and reproduction has also been extensively explored. Studies have examined the effects of animal manures (Pulikeshi *et al.*, 2001), various food sources (Suthar et al., 2008), and specific organic wastes (Biruntha et al., 2013; Sadia et al., 2020) on different species. Comparative analyses of life cycles and reproductive traits across species like *Lampito mauritii, Metaphire posthuma, Perionyx ceylanensis, and Aporrectodea trapezoides* have yielded valuable information for soil management and vermicomposting.

Objective

The life cycle dynamics of numerous tropical earthworm species are yet to be fully characterized. Despite existing life cycle studies, the indigenous species *Drawida modesta* remains largely unstudied. Therefore, this research endeavors to investigate and document specific aspects of the life cycle of this species.

Materials and Methods

Description of *Drawida modesta*:

Drawida modesta is characterized by the following morphometric and anatomical features: Somatic dimensions include a length of 75-100 mm and a width of 4-5 mm, with a segment count of approximately 209. The integument exhibits a cream-brown coloration. Diagnostic features include the absence of dorsal pores and an epilobic prostomium. Nephridiopores are aligned with the d setae. Setae are closely paired, with intersetal relationships of aa=bc and dd approximating half the body circumference or slightly exceeding it. The male pore complex is located on the 10/11 intersegmental region, featuring prominent anterior and posterior labia positioned just lateral to the b setae, accompanied by transverse groove-like depressions both anterior and posterior to the pores. The midventral regions of segments 10 and 11 display a slight depression. Female pores are present. Spermathecal pore apertures are positioned slightly lateral to the b setae, each located on a slightly elevated, transversely oval, flat papilla. Internal anatomical features include thickened septa from 5/6 to 8/9, two gizzards located in segments 12 and 13, and testis sacs that extend into both segments 9 and 10.

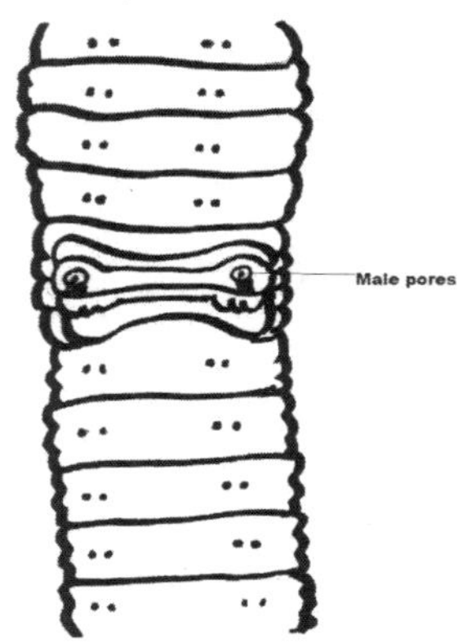

Fig: 1 *Drawida modesta* Rao, 1921 (ventral view)

Culture technique:

Mature, clitellate individuals of *Drawida modesta* were collected from a banana field through manual excavation and hand-picking. Taxonomic identification was performed based on established morphological characteristics and taxonomic keys. The collected earthworms, along with their native soil, were transported to the laboratory for subsequent rearing and experimentation.

For controlled cultivation, standardized rearing beds were established using plastic containers (18 cm diameter x 8 cm) with perforated lids to ensure adequate aeration. The culture medium consisted of a 1:1 mixture of soil, 25 grams of 15-day-old cow dung, and air-dried banana leaves. Moisture levels were maintained close to field soil moisture by intermittent water spraying. Before introducing the clitellate worms, the beds were allowed to cure for several days. Daily atmospheric temperature was recorded, and three replicate cultures were maintained. Observations included the size, color, and shape of deposited cocoons. Continuous monitoring of the culture beds was conducted. At each sampling interval, the culture medium was replaced with fresh material to prevent food availability from becoming a limiting factor.

Plate 1: Culturing boxes for experiment

Cocoon studies:

Weekly examinations of each culture container were conducted, involving the manual sorting and weighing of adult *Drawida modesta* individuals. Cocoons were extracted from the culture media using a gentle wet sieving technique with a 0.05 mm mesh sieve, and the number of cocoons produced per individual was quantified. The size and weight of the cocoons were subsequently measured, with the cocoons being washed prior to weighing to remove any adhering debris. To ensure that food availability did not become a limiting factor, the culture media was replaced with fresh material at each sampling interval.

Plate 2: Adult *Drawida modesta* earthworms

Incubation of cocoons:

To determine the incubation period and hatchling characteristics, isolated *Drawida modesta* cocoons were individually incubated on water-saturated cotton (80% moisture) at a consistent temperature mirroring the earthworms' production environment. During incubation, cocoons were regularly monitored for color changes and hatchling emergence. The total number of hatchlings per cocoon, their initial weight, size, and emergence point were recorded, enabling the calculation of hatchability and incubation duration. The incubation period was calculated as the time elapsed from cocoon collection to the first hatchling appearance, adjusted by adding half the time interval between collection and the preceding inspection of the culture medium, adhering to the methodology established by Butt (1993, 1997).

Results

Drawida modesta exhibits a semi-continuous reproductive strategy, demonstrating peak reproductive activity during the monsoon and post-monsoon seasons. In controlled artificial bed environments, the average annual cocoon production was quantified at 27 cocoons per individual. Freshly deposited cocoons were characterized as oval, cream-white, and semi-transparent, with an average weight of 0.03 grams. Cocoons were typically deposited within the upper 10-15 cm of the substrate. The mean incubation period was determined to be 26 days, with no significant seasonal variation observed. Each cocoon yielded an average of 2-3 hatchlings, indicating polyembryony. Newly emerged hatchlings were slender and pinkish, with a mean length of 0.5-1.8 mm and an average weight of 0.05 grams. Following clitellum formation, the initiation of cocoon deposition occurred within a mean period of 5-8 days.

The incubation period of *Drawida modesta* cocoons was determined to be 26 days, with notable seasonal variations observed. Polyembryony, evidenced by the presence of multiple hatchlings per cocoon, was documented. Newly hatched individuals were characterized as slender, soft, and whitish, gradually transitioning to a harder, brownish coloration. The time required for hatchlings to reach reproductive maturity, as indicated by clitellum development, was 70 days. The clitellar region exhibited a slightly pinkish hue compared to the rest of the body. *D. modesta* displays a semi-continuous breeding pattern, with cocoon production occurring from April to October. During the winter months (November to February), the species enters a state of quiescence, characterized by the cessation of reproductive activity.

Table 1: Ecological category, habitat and size of *Drawida modesta*

Time course of cocoon production	Semi-continuous
Appearance of the clitellum (days)	27- 32 days
Colour	Lightly pigmented dorsally
Length	42-50 mm
Feeding habit	Phytophagous
Site of collection	Banana field

Vertical distribution	10-15 cm
Ecological category	Top soil endogeic species
Weight	3.48 mg
Mode of reproduction	Cross fertilization
Incubation period	26 days
Rate of cocoon production	27 cocoon/worm/year
Number of hatchlings from one cocoon	2, sometimes more than 3

A

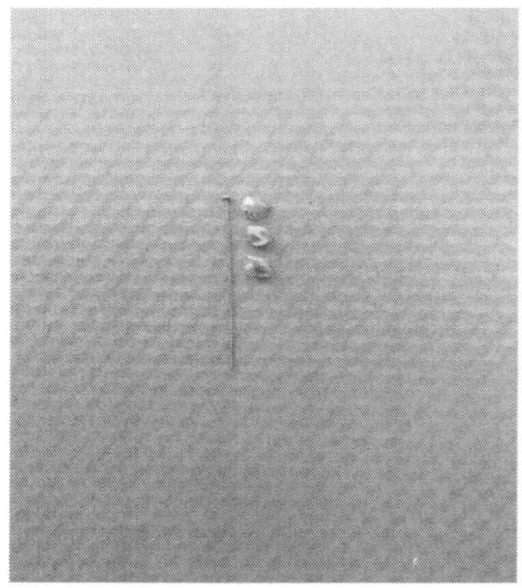

B

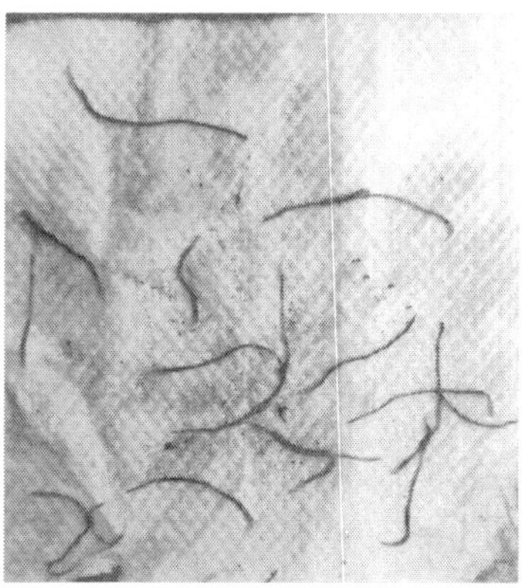

A

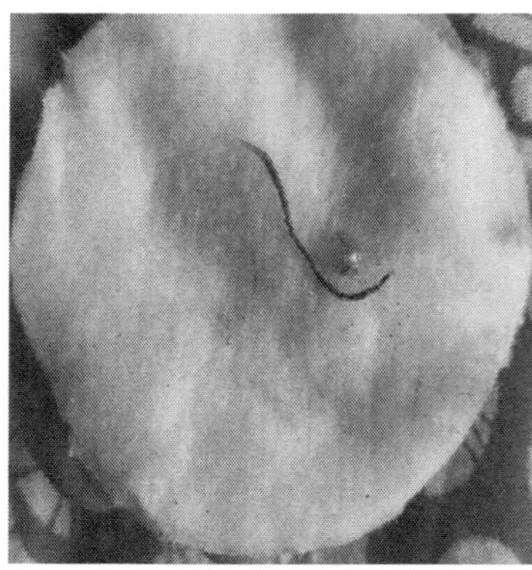

B

Plate 3: : *Drawida modesta*: A. Freshly laid cocoon in culturing box; B. Change in colour and shape of cocoon after Hatching; C. Cocoon with juvenile, D. Juvenile after few days of Hatching

Fecundity

The fecundity of *Drawida modesta* was determined by quantifying the number of cocoons produced per adult individual over a one-year period, revealing an average of 27 cocoons per adult. Seasonal analysis of cocoon production demonstrated a clear correlation with temperature and precipitation patterns. Peak cocoon production occurred during the summer months, with a moderate reduction observed during the monsoon season, and a significant decline during winter. This seasonal trend in cocoon production was reflected in the calculated fecundity, with the highest cocoon production rates observed between March and June, followed by a decrease from July to October, and the lowest rates recorded from November to February.

Regeneration

Following anterior amputation, *Drawida modesta* exhibits a distinct defensive behavior characterized by immediate body coiling into a tight, knot-like structure. Conversely, posterior amputation results in the exudation of a yellow secretion. Wound healing occurs within five to seven days post-amputation. Notably, D. modesta demonstrates the capacity for posterior regeneration, but lacks anterior regeneration. Regenerated segments initially exhibit a light pink hue, with pigmentation gradually intensifying over time. After 38 days, approximately three to four segments are regenerated in the posterior region.

Table 2: Rate of cocoon production by *Drawida modesta* earthworm species (Monthly study). Mean ± standard error

Month Tempera-ture	Jan-uary 19-22° C	Febr-uary 25.5-26.5° C	March 30.5-32.5°C	April 34.4-35.5° C	May 32.5-33.2°C	June 32-33.7 C	July 30.1-32.9° C	August 30.5-31.5° C	Sept-ember 29-30°C	Octo-ber 26.5-27°C	Nove-mber 23.5-25° C	Dece-mber 22-23° C
Drawida modesta (per worm)	00.00± 0.00	02.00± 0.00	02.20± 0.30	04.70± 0.23	2.90± 0.20	02.02± 0.00	3.25± 0.20	1.70± 0.20	02.00± 0.25	02.15± 0.30	01.85± 0.30	2.25± 0.25

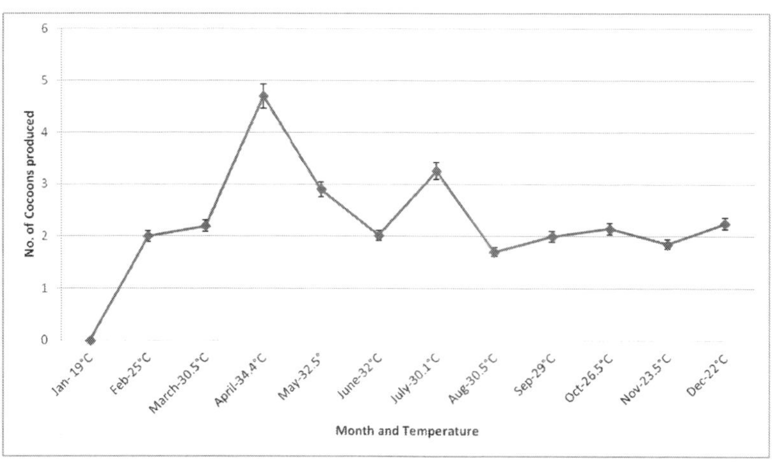

Graph 1: Number of cocoons produced per worm by Drawida modesta species

Discussion

According to Satchell (1967) the growth and reproduction in earthworms is associated with their ecological strategies. Long growth phase and low reproductive rate is the one of the reproductive strategy in endogeic earthworms whereas the short growth phase and high rate of reproduction can be seen in epigeic earthworms. This type of variation is adapted by the earthworms to increase the survival chances in favorable and in unfavorable conditions respectively. The endogeic are soil living generally found in a more stable, undisturbed environment and their exposure to adverse conditions and predators is very limited. Because of this reason, endogeic species show slow growth rate and produce less number of young ones. Whereas the epigeic earthworms are more prone to exposure to adverse conditions and predator is quiet common in epigeic species as they live above the top soil ground their survival is in a very uncertain in nature. Due to this reason these species are have to be very active, feed continuously on rich in organic matter, grow fast and produce large number of cocoons.

Drawida modesta being endogeic species showed similar type of adaptability by exhibiting slow growth and low reproductive rate. Lee (1985) showed the correlation between the high risk mortality and earthworm species with its high rate of cocoon production. Cocoon production and time of maturation of cocoons is always species specific. Population density, age, structure and external abiotic factors especially soil temperature, moisture and energy content of the available food play vital role in reproduction stage of earthworms (Lee, 1985).

The earthworms are Hermaphrodites and reproduce cocoons by means of cross fertilization, self fertilization, or through parthenogenesis. In our study it was observed that *Drawida modesta* being clitellated could not reproduce through parthenogenesis but it reproduces through cross fertilization our observations in line with the findings of Chaudhuri and Subhalaxmi Bhattacharjee (2011) who reported about the life cycle of *Drawida assamensis, Drawida papillifer papillifer.* According to Lavelle et al. (1998) native species produce viable cocoons only if fertilization occurs.

Lavelle *et al.* (1998) and Barois et al. (1999) proposed a relationship between reproductive strategies and ecological categories in tropical earthworms. According to them there are four categories of earthworms. Category 1: large native endogeic and anecic species with low fecundity (0.5 to 3.1 cocoons per adult/per year) with one hatchling per cocoon. Category 2 medium-sized meso humic endogeic species with moderate fecundity (1.3–4.5 cocoons adult/per year) with more than 2 hatchling per cocoon; Category 3: small mainly poly humic endogeic species with intermediate fecundity (10 to 68 cocoons per adult/per year) and usually one hatchling per cocoon; and Category 4: generally small, mainly exotic and epigeic species with very high fecundity (50 to 350 cocoons per adult/per year) and up to three hatchlings per cocoon. From our study it is observed that *Drawida modesta* earthworm species belongs to category 2 with low count of cocoons.

The mean incubation period for *Drawida modesta* was 26±1.5 days, which was very similar to that reported for *Drawida nepalensis* 28.7 days by Kaushal et al. (1995). *D. rubida* required 36.5 days as reported by Elvira *et al.* (1996). While 43 – 90 days for *D. venseta* by Fayolle *et al.* (1997).

Edwards and Lofty (1977) reported that decreased supply of organic matter in the food content gradually decreases the earthworm population. Decrease in the availability of food in the airable land was associated with decrease in number of cocoons (Evans and Guild 1948) the present investigation has similar findings with the above as 50% of earthworm population reduced by the end of the experiment.

Conclusion

Drawida modesta exhibits a semi-continuous reproductive strategy characterized by moderate fecundity, with seasonal variations in cocoon production. Specifically, a notable reduction or cessation of reproductive activity is observed during the winter months, while elevated cocoon production occurs during the summer and monsoon seasons, peaking in April and July. The species demonstrates a moderate regenerative capacity. As a topsoil endogeic organism, *D. modesta* presents potential applications in vermiculture. Its capacity to facilitate the conversion of organic waste into valuable vermicompost and vermiwash underscores its potential contribution to sustainable waste management and soil fertility enhancement. Successful implementation of vermicomposting practices utilizing *D. modesta*, driven by community-level engagement, holds promise for the reclamation and revitalization of degraded land resources.

Acknowledgement

The authors express their sincere gratitude to retired Professor Dr. Shankerappa S. Hatti of Government First Grade College, Kalaburgi, for his invaluable support and guidance throughout the conduct of this research.

References

Bahl, K. N. (1922). On the development of the enteronephric type of nephridial system found in Indian earthworms of the genus *Pheretima*. *Quarterly Journal of Microscopical Science*, 2-66: 49-103.

Bahl, K. N. (1927). On the reproductive processes of earthworms. Part 1. The process of copulation and exchange of sperms in *Eutyphoeus waltoni* Mich. *Quarterly Journal of Microscopy*, 2(71): 479-502.

Barois, I., Lavelle, P., Brossard, M., Tondoh, J., Angeles Martinez, M., Rossi, J. P., Senapati, B. K., Angeles, A., Fragoso, C., Jimenez, J. J., Decaëns, T.,

Lattaud, C., Kanyonyo, J., Blanchart, E., Chapuis, L., Brown, G. and Moreno, A. (1999). Ecology of earthworm species with large environmental tolerance and/or extended distribution. pp. 57-84.

Biruntha, M. (2013). Growth and reproduction of *Perionyx excavatus* in different organic wastes. *International Journal of Current Microbiology and Applied Science,* 2: 28-35.

Bouche, M. B. (1983). The establishment of earthworm communities. In: Satchell, J. E. (ed.) *Earthworm ecology*: From Darwin to vermiculture. Chapman and Hall, London, pp. 431–448.

Butt, K. R. (1993). Reproduction and growth of three deep-burrowing earthworms (Lumbricidae) in laboratory culture in order to assess production for soil restoration. *Biol. Fert. Soils,* (16): 135–138.

Butt, K. R. (1997). Interactions between selected earthworm species: a preliminary, laboratory based study.

Chaudhuri, P. S. and Bhattacharjee, S. (2011). Reproductive biology of eight tropical earthworm species of rubber plantation in Tripura, India. *Tropical Ecology,* 52(1): 49-60.

Edwards, C. A. and Lofty, J. R. (1972) *Biology of earthworms.* Chapman and Hall, London, pp. 283.

Edwards, C. A. and Lofty, J. R. (1977). Biology of earthworms. Chapman and Hall, New York, pp. 333.

Edwards, C. A. and Lofty, J. R. (1972). *Biology of Earthworms.* Chapman and Hall, London.

Elvira, C., Dominguez, J. and Briones, M. J. I. (1996). Growth and reproduction of *Eisenia andrei* and *Eisenia fetida* (Oligochaeta, Lumbricidae) in different organic residues. *Pedobiologia,* 40: 377-384.

Evans, A. C. and Mc Guild, W. J. (1948). Studies on the relationships between earthworms and soil fertility. IV. On the life cycles of some British Lumbricidae. *Annals of Applied Biology,* 35: 471-484.

Fayolle, L., Michaud, H., Cluzeau, D. and Stawieck J. (1997). Influence of temperature and food source on the life cycle of the earthworm *Dendrobaena veneta* (oligochaeta). *Soil Biology and Biochemistry,* 29: 747 – 50.

Fragoso, C. and Lavelle, P. (1992). Earthworm communities of tropical rain forests. Soil Biol. Biochem., 24(12): 1397-1408.

Julka, J. M. (1988). *The fauna of India and adjacent countries. Megadrile Oligochaeta (Earthworms)* Haplotaxida: Lumbricina: Megascolecoidea: Octochartidae. Zoological Survey of India, Calcutta. xiv+, 400.

Kale, R. D. and Bano, K. (1991). Time and space relative population growth of *Eudrilus eugeniae.* In: Veeresh, G. K., Rajagopal, D. and Viraktamath, C. A., eds, Advances in management and conservation of soil fauna. New Delhi: Oxford and IBH, pp. 657–664.

Kale, R. D. and Bano, K. (1994). Laboratory studies on age specific survival and fecundity of *Eudrilus eugeniae. Mitt. Hamb. Zool. Mus. Inst.* (89): 139-148.

Kale, R. D., Bano, K. and Krishnamoorthy, R. W. (1982). Potential of *Perionyx excavatus* for utilizing organic wastes. *Pedobiologia,* 23: 419–426.

Kaplan, D. L., Hartenstein, R. and Neuhauser, E.F. (1980). Coprophagic relations among the earthworms *Eisenia foetida. Eudrilus eugeniae* and *Amynthas sp. Pedobiol.,* 20: 74 - 84.

Kaushal, B. R. and Bisht, S. P. S. (1992). Growth and cocoon production of *Drawida nepalensis* (Oligochaeta). *Biol. Fertil. Soils,* 14: 205–212.

Kaushal, B. R., Kalia, S. and Bisht, S. P. S. (1995). Growth and cocoon production by the earthworm *Drawida nepalensis* (Oligochaeta: Moniligastridae) in oak and pine litter. *Pedobiologia,* 39: 417-425.

Lavelle, P., Barois, I., Blanchart, E., Brown, G., Brussaard, L., Decaëns, T., Fragoso, C., Jimenez, J. J., Kajondo, K. K., Ángeles Martinez, M. D. L., Moreno, A., Pashanasi B., Senapati B. K. and Villenave, C. (1998). Earthworms as a resource in tropical agroecosystems. *Nature Resources,* 34: 26-41.

Lavelle, P., Lattaud, C., Trigo, D. and Barois, I. (1995). Mutualism and biodiversity in soils. *Plant and Soil.* 170:23-33.

Lee, K. E. (1985). Earthworms their ecology and relationships with soils and land use. *Academic Press, Australia.* pp. 411.

Piearce, T. G., Oates, K. and Carruthers, W. J. (1990). A fossil earthworm embryo (Oligochaeta) from beneath a late Bronze Age midden at Potterna, Wiltshire, UK., *J. Zool. Land.* 220: 537-542.

Pulikeshi, M. B, Amoji, S. D, Shagoti, U. M. and Biradar, V. A. (2001). Influence of animal manures on the biology of temperate earthworm, *Eisenia fetida* in tropical semiarid climate. *J. Environ. Biol.,* 22(2): 113-8.

Reinecke, A. J. and Krief, J. R. (1981). The influence of constant and diurnally fluctuating temperatures on the cocoon production, hatching time and number of hatchlings of Ei*senia foetida* (Lumbricidae, Oligochaeta). In: Appelholf M (ed). Role of earthworms in the stabilization of organic residues beech leaf, Kalamazoo, Mich. pp: 167 – 77.

Reinecke, A. J. and Kriel, J. R. (1981). Influence of temperature on the reproduction of the earthworm Eisenia foetida (Oligochaeta), *South African Journal of Zoology.* 16:96-100.

Sadia Masuma., Hossain Md., Islam Md., Akter Taslima., Dinesh Chandra and Shaha. (2020). Growth and reproduction performances of earthworm (*Perionyx excavatus*) fed with different organic waste materials, 7: 331-337.

Satchell, J. E. (1967). Lumbricidae In: *Soil Biology*. (A. Burges and F. Raw, Eds). *Academic Press*. London, 259-322.

Stephenson, J. (1930). *The Oligochaeta*. Oxford University Press, Oxford, U.K.

Suthar Surindra (2008). Influence of different food sources on growth and reproduction performance of composting epigeics: *Eudrilus eugeniae, Perionyx excavatus* and *Perionyx sansibaricus*. *Applied Ecology and Environmental Research,* 5: 1-10.

Turbe, A., Toni, A. D., Benito, P., Lavelle, P., Lavelle, P., Ruiz, N., Van der Putten, W. H, Labouze, E. and Mudgal, S. (2010). Soil biodiversity: functions, threats and tools for policy makers. Bio Intelligence Service, IRD and NIOO, Report for European Commission (DG Environment).

Viljoen, S. A. and Reinecke, A. J. (1989). The life cycle of the African night crawler, *Eudrilus eugeniae* (Oligochaeta); *S. Afr. J. Zool.*, 24: 27–32.

Fig: 1 Rosmarinus *officinalis* (Pg 47)

Fig: 3 Withania *somnifera* (Pg 48)

Fig: 5 *Emblica officinalis* (Pg 50)

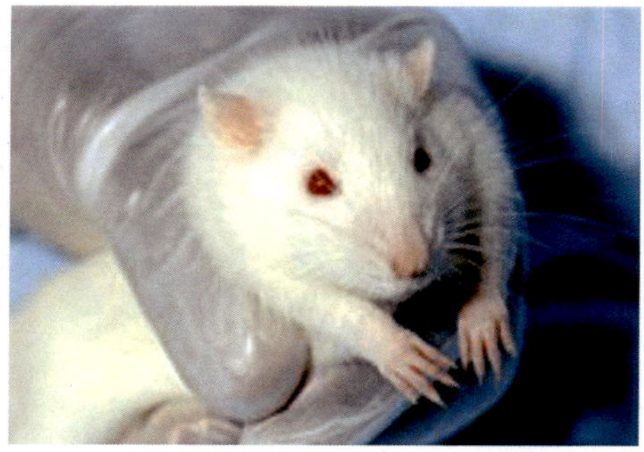

Fig: 7 Sprague dawley (female) (Pg 52)

Fig: 8 Acclimatization of rat (Pg 52)

2.a. Riverine area (Pg 91)

2.b. Cropland area (Pg 91)

2.c. Urban area (Pg 92)

Fig: 2 Photographs of different sites (Pg 92)

Fig: 3 Merops orientalis birds photographs collected in study area (Pg 93)

Fig: 3 Study area: Honnali (Pg 102)

Fig: 4 Study area: Honnali (Pg 102)

Fig: 5 *Pteropus gigenteus* in the branches of *Ficus* tree (Pg 102)

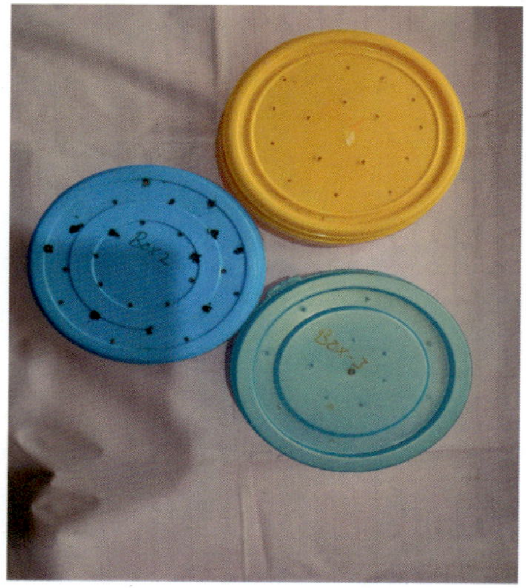

Plate 1: Culturing boxes for experiment (Pg 116)

Plate 2: Adult *Drawida modesta* earthworms (Pg 116)

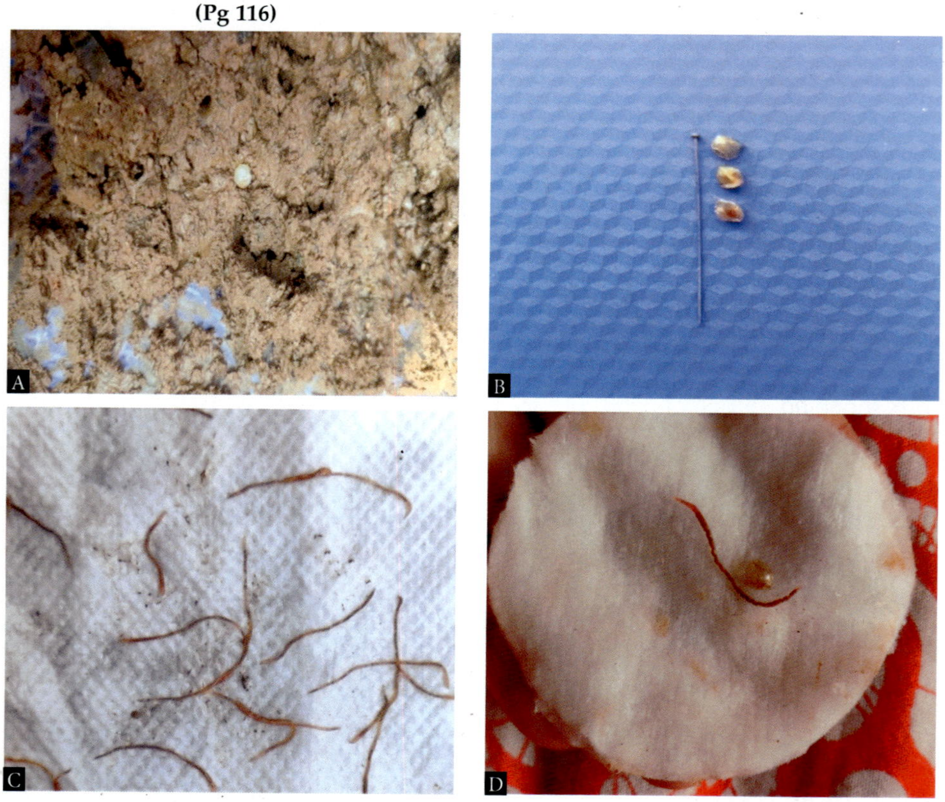

Plate 3: : *Drawida modesta*: A. Freshly laid cocoon in culturing box; B. Change in colour and shape of cocoon after Hatching; C. Cocoon with juvenile, D. Juvenile after few days of Hatching (Pg 118)

CHAPTER 11

BIO-INSPIRED OF SILVER NANOPARTICLES USING PHYTOEXTRACT: A GREEN TOOL FOR BIOMEDICINE

Bheemanagouda N. P.[1,2*], Chandappagoud P.[3], Prashanth Konkal[4*], Arun K. S.[5], Taranath T. C.[2] and Channabasava[4]

[1]*Department of Botany Smt. ASM College for Women Ballari, Karnataka, India*
[2]*P. G. Department of Studies in Botany, Environmental Biology Laboratory,*
[4]*Karnatak University, Dharwad-580003. Karnataka, India.*

[3]*Department of Obstetrics and Gynecology*
Sri Sanganabasava Scanning Centre and Maternity Hospital Kembhavi, Karnataka, India
Corresponding Author: bnpatil089@gmail.com, prashanthkonkal@gmail.com

Abstract:

The preparatipon of biogenic silver nanoparticles (AgNPs) using the various secondary metabolites namely terpenoids, polyphenols, sugars, alkaloids, phenolic acids, and proteins, play an important role in the bioreduction and biocapping of metal ions, leading to the formation of AgNPs. These secondary metabolites are non toxic chemicals, environmentally benign solvents and renewable materials are some of the key issues that merit important consideration in a green synthetic strategy. The use of higher plants in the synthesis of AgNPs is quite a novel method as it is cost effective, environmental friendly and easily scaled up for large scale synthesis. The biomamatic silver nanoparticles were characterized by using UV-

Vis spectroscopy (UV-Vis), Atomic force microscopy (AFM), X-Ray diffraction (XRD), Fourier transform infrared (FTIR). They offer better built, long lasting, cleaner, safer and smatter products for using antibacterial, antifungal, antiviral, antioxidant, anticancer, neuroprotective activity and other medical activity. This work highlights the properties of AgNPs, how miniaturization cause change in the properties of AgNPs, significance of physicochemical factors governing the synthesis, biomechanical properties of AgNPs synthesized by higher plants and their biomedical applications.

Keywords: Angiosperms, AgNPs, Bioreduction, ROS, Applications

Introduction

It has been positioned as the 47th element in the periodic table, having an atomic weight of 107.8 and two natural isotopes 106.90 Ag and 108.90 Ag with abundance of 52 and 48% where as the colloidal silver is of particular interest because of distinctive properties such as good conductivity, chemical stability, catalytic and antibacterial activity [1]. The medicinal and preservative properties of silver have been known for over 2,000 years. It is a rare, but naturally available, slightly harder than gold and very ductile and malleable. AgNPs of many different shapes (spherical, rod-shaped, truncated, triangular nanoplates) were developed by various synthetic methods. Truncated triangular silver nanoplates were found to show the strongest anti-bacterial activity. The AgNPs have excellent antimicrobial property compared to other salts due to their extremely large surface area, which gives good contact with microorganisms. Silver ions and nanoparticles are highly toxic and hazardous to microorganisms. AgNPs have many applications like they might be used as spectrally selective coatings for solar energy absorption and intercalation material for electrical batteries, as optical receptors, as catalysts in chemical reactions in biolabelling, and as antimicrobials [2-4]. Synthesis of AgNPs using non vascular plant [5-7], fungal [8,9], Bacteria extracts [10,11], cytotoxic effects of AgNPs [12]. Current research in inorganic nanomaterials having good antimicrobial properties has opened a new era in pharmaceutical industries [13-16]. Silver is the metal of choice as they hold the promise to kill microbes effectively. AgNPs have been recently known to be a promising antimicrobial agent that acts on a broad range of target sites both extra cellular as well as intra-cellular. The resistance conferred by bacteria to silver is determined by the 'sil' gene in plasmids [17].

AgNPs have exhibited antimicrobial effect on *Staphylococcus aureus* and *Escherichia coli* [18]. NP take advantages of the oligodynamic effect that silver has on microbes [19]. Green synthesis of small AgNPs using geraniol and its cytotoxicity against cancer cell line Fibrosarcoma-Wehi 164 [20]. Whereas the cell proliferation was evaluated by a modified crystal violet colorimetric assay [21].

Green synthesis, antimicrobial and cytotoxic effects of AgNPs using *Eucalyptus chapmaniana* leaves extract [22].

Objective

The life cycle dynamics of numerous tropical earthworm species are yet to be fully characterized. Despite existing life cycle studies, the indigenous species *Drawida modesta* remains largely unstudied. Therefore, this research endeavors to investigate and document specific aspects of the life cycle of this species.

2. Characterization of nanoparticles using the following methods

2.1 UV-Visible Spectroscopy

Ultraviolet-visible spectrophotometer (UV-Vis) refers to absorption spectroscopy in the ultraviolet-visible spectral region. This means that it utilizes the light in the visible and adjacent (near-UV and near-infrared [NIR]) ranges. In the visible range the absorption or reflectance directly affects the perceived color of the involved chemicals. The molecules undergo electronic transitions in the electromagnetic spectrum region [24]. UV spectroscopy is type of absorption spectroscopy where the molecule absorbs the light in the ultra-violet region (200-400 nm.). The electrons are excited from the ground state to the higher energy state by the absorption of the ultra-violet radiations. Working on the Beer-Lambert law principal. It states that "when a beam of monochromatic light is passed through a solution of an absorbing substance, the rate of decrease of intensity of radiation with thickness of the absorbing solution is proportional to the incident radiation as well as the concentration of the solution".

2.2 Fourier Transform Infra-Red (FT-IR) Spectroscopy

Fourier transform infrared spectroscopy specific information about the vibration, rotation of the chemical bonding and molecular structure is given by Fourier transform infrared spectroscopy. This information is useful in analyzing certain in-organic and organic materials. An infrared spectrum represents a fingerprint of a sample with absorption peaks. These peaks correspond to the frequencies of the vibrations between the bonds of the atoms thus making up the material. Different material has a unique combination of atoms, and hence no two compounds produce the same infrared spectrum. Therefore, a positive identification (qualitative analysis) is provided by infrared spectroscopy of every different kind of material. In addition, a direct indication of the amount of material present is detected by the size of the peaks in the spectrum. Infrared spectroscopy is an excellent tool for quantitative analysis with modern software algorithms. The IR region is divided into three smaller areas: near - IR (400 - 10 cm-1), mid - IR (4000 - 400 cm^{-1}), and far - IR (14000 – 4000 cm^{-1}).

2.3 X-ray diffraction (XRD)

The most common and efficient method for the determination of structure, crystallinity and material identififcation is the X - ray diffraction technique. XRD is very good method to detect whether a resultant material is amorphous or crystalline in nature. Crystalline phases can be identified by comparing the interplanar distance 'd' values which are obtained from XRD data, with the data in Joint Committee on Powder Diffraction Standards (JCPDS).

2.4 Energy Dispersive X-ray Spectroscopy (EDS) Analysis

Energy Dispersive X-Ray Spectroscopy (EDS or EDX) is used in conjunction with scanning electron microscopy, which is in fact a chemical microanalysis technique. In this technique the sample is bombarded by the SEM's electron beam. Due to the bombardment, the electrons are ejected from the atoms comprising the sample's surface. The vacancies created by the ejected electrons are filled by electrons from a higher state. To balance the energy difference between the two electrons' states an x-ray is emitted. The x-ray energy is characteristic of the element from which it was emitted.

2.5 Atomic Force Microscope (AFM)

The principle of operation of the AFM is very similar with that of stylus profilometer – a sharp cantiever tip interacts with the sample surface sensing the local force between the molecules of the tip and the sample surface. This instrument is not a 'conventionqal microscope' that collects and focuses light. The word microscope has been associated with this instrument because it is able to measure microscopic features of the samples. The most characteristics property of the AFM is that the images are acquired by 'feeling' the sample surface without using light. In this way, not only the sample topography can be recorded with good resolution, but also the material characteristics and the strength of interaction between the sample surface and the cantilever tip.

2.6 High Resolution Transmission Electron Microscope

Basic principle of HR-TEM is quite similar to the other optical counterparts, the optical microscope. The major difference in HR-TEM is that, to "image" and achieve information about the structure and composition of the specimen, a focused beam of electrons is used instead of light. An electron source, the "Gun" is used to produce a stream of electrons. By using a positive electrical potential these electrons are accelerated towards the specimen. Metal apertures and magnetic lenses called "condenser lenses" focus this stream into a thin, focused, monochromatic beam. A part of the beam gets transmitted through the specimen as beam strikes the specimen. A set of lenses called "objective lenses" again focus this portion of the beam into an image. Through the "intermediate and projector lenses", this image is then fed down the column. Then depending upon the magnification set the image is enlarged. The image is produced on a phosphor image screen. The light is engendered as the image strikes the screen which later enables the user to see the image. In the image those areas of the sample which are thinner or less dense more electrons were transmitted and appear as the lighter areas and those areas of the sample which are thicker or denser region where fewer electrons were transmitted appear as darker areas of the image.

3. Phytosynthesis of Silver Nanoparticles

Several microorganisms are used for the synthesis of AgNPs such as organisims like bacteria, fungi andyeasts have come up as nanofactories synthesizing metalNPs of Ag. However, use of plants for the fabricationof NPs has drawn attention of workers because of itsrapid, economical, eco-friendly protocol and it providesa single step technique for the biosynthesis process. Antimicrobial activities of AgNPs One of the basic elements that make up our planet is silver. Silver, a naturally occurring element is slightly harder than gold and is malleable and ductile. Pure silver has very low contact resistance and highest thermal and electrical conductivity compared to all the metals. Metallic salts such as AgNO3 and silver chloride are soluble in water (WHO, 2002) while metallic silver is insoluble in water. Silver in its metallic form is used in coinage, as fungicides and in surgical prosthesis and splints. Soluble silver salts, were used in treating epilepsy, nicotine addition, mental illness, and gastroenteritis. It is also used in the treatment of infectious diseases such as *gonorrhoea* and *syphilis* [25]. Toxicity of silver is dependent in the presence of free silver ions in the environment. But investigations by World Health Organization (WHO), 2002 have revealed that these concentrations of silver ions are very low and are non-toxic to the environment. Intake of silver is estimated at 70-90µg/day. According to studies silver in any form is non-toxic to the cardiovascular system, nervous system, immune system or reproductive system nor is it considered to be a carcinogenic substance [26], hence silver is considered to be non-toxic [27]. The table 1 show the different nanoparticles synthesized from the plant extract and their activity.

Table 1: Various plant extract used for the green synthesis of AgNPs and their biomedical activity

Sl. No	Name of the plant	Part used	Size and shape of nanoparticles	Biomedical applications	Ref.
1.	*Limonia acidissima L.*	Leaf	21-42 spherical	Antimicrobial	[28]
2.	*Cinnamon zeylanicum* Blume	Bark	30-40nm, Quasi-spherical and small, rod-shaped.	Antibacterial activity	[29]
3.	*Musa* spp.L.	Peel	Nanocrystallites	Antibacterial and antifungal activity	[30]

4.	*Curcuma longa* L.	Tuber powder	21 and 30 nm, Quasi-spherical, triangular and small rod-shaped	Antibacterial activity	[31]
5.	*Piper nigrum* L.	Leaves	20–50nm , crystalline silver nanoparticles	Antibacterial activity	[32]
6.	*Sesuvium portulacastrum* L.	Leaves	5 to 20 nm, Spherical	Antibacterial and antifungal activity	[33]
7.	*Citrullus colocynthis* (L.) Schrad .	Leaves	31 nm, Spherical	Anticancer drug	[34]
8.	*Garcinia mangostana* L.	Leaves	35 nm, Spherical	Antibacterial activity	[35]
9.	*Annona squamosa* L.	Peel	35 ± 2 nm, irregular spherical	Acaricidal, insecticidal, and larvicidal activity	[36]
10.	*Ficus benghalensis* L.	Leaves	16 nm, spherical	Antibacterial activity	[37]
11.	*Piper longum* L.	Leaves	17.6– 41 nm , Spherical	Cytotoxic activity against Hep-2 cell line	[38]
12.	*Terminalia chebula* Retz.	Fruits	Less than 100 nm, ,Pentagons, spherical and triangular	Antibacterial activity	[39]
13.	*Paederia foetida* L.	Leaf	24nm, spherical	Antibacterial activity	[40]
14.	*Saraca indica* (Roxb.) Wilde	Bark	5 to 50 nm, spherical	Analgesic activity	[41]
15.	*Murraya koenigii* (L.) Sprengel	Leaf	57 nm, spherical	Antibacterial activity	[42]
16.	*Cocculus hirsutus* (L.) Diels.	Leaf	139nm, spherical	Antibacterial activity	[43]
17.	*Eriobotrya japonica* (Thunb.) Lindl.	Leaf	18 nm, Spherical	Antibacterial activity	[44]
18.	*Annona reticulate* L.	Leaf	23.84 to 50.54 nm, spherical, rod, and triangular	Antibacterial and antifungal	[45]
19	*Rubus idaeus* L.	Leaf	Below 100nm, spherical and square shaped	Antibacterial activity	[46]

20.	*Cymbopogan citratus* (DC) stapf .	Leaf	15 to 65 nm, cuboidal, and rectangular	Antibacterial and antifungal	[47]
21.	*Tinospora cordifolia* (Thunb.) Miers	Stem	4-20nm, spherical	Antibacterial activity	[48]
22.	*Punica granatum* L.	Fruit peel	30-40nm, spherical	Anticancerous activity	[49]
23.	*Abrus precatorius* L.	Leaf	14 nm, Spherical	Antibacterial and antifungal activity	[50]
24.	*Leptadenia reticulate* (Retz.) W&A Shrub	Leaf	50-70nm, spherical	Antimicrobial and antioxidents	[51]
25.	*Dodonaea viscose* Jacq.	Leaf	29nm, spherical	Antimicrobial	[52]
26.	*Tribulus terrestris* L.	Fruit	16-28nm, Spherical	Antibacterial	[53]
27.	*Iresine herbstii* Hook. ex Lindl.	Leaf	44-64nm, Cubic	Antimicrobial & antioxidant	[54]
28.	*Jatropha curcas* L.	Latex	>263 nm, Round shape	Antimicrobial	[55]
29.	*Solanum trilobatum* L.	Leaves	15-20 nm, Cubic, hexagonal	Antidandruff, antimicrobial	[56]
30.	*Ceratonia siliqua* L.	Leaf	5-40 nm, spherical	Antibacterial	[57]
31.	*Jatropha gossypifolia* L.	Latex	>263 nm, Round shape	Antimicrobial	[58]
32.	Zanthoxylum acanthopodium DC	Fruit	55.67, Cubic	Antimicrobial Activity	[59]
33.	*Euphorbia milii* Des Moul.	Latex	>263 nm, Round shape	Antimicrobial	[60]
34.	*Citrus limon*(L.) Burm.f.	Leaves	>100 nm., Multi-shaped	Antimicrobial	[61]
35.	*Catharanthus roseus* (L.) G.Don	Leaves	29-31nm	Antimicrobial	[62]
36.	*Origanum vulgare* L.	Leaves	63-85nm, Spherical	Antibacterial and Anticancer activity	[63]
37.	*Curcuma longa* L.	Rhizome	21-30nm, Spherical	Bactericidal activity	[64]
38.	*Garcinia mangostana* L.	Leaf	6-57nm,	Antimicrobial	[65]

39.	*Ocimum tenuiflorum* L.	Leaf	25-40nm, Spherical	Antibacterial	[66]
40.	*Morus* spp.L.	Leaf	20-40nm, cubical and rela-tively uniform	Antibacterial	[67]
41.	*Vernonia amygdalina* Delile	Plant	50-70nm, spherical and triangular	Antimicrobial	[68]
42.	*Ocimum sanctum* L.	Leaf	18nm, Spherical	Antibacterial	[69]
43.	*Cassia auriculata* L.	Flower	10-40 nm, spherical	antioxidant activities	[70]
44.	*Albizia adianthifolia* (Schumach.)W.F. Wight	Leaf	4–35 nm, spherical	A549 lung cell line activity	[71]
45.	*Podophyllum hexandrum* Royle	Leaf	12-40nm, spherical	Cytotoxicity and Caspase-mediated apoptotic	[72]
46.	*Leucas aspera* (Willd.) Link	Bark	29–45 nm, spherical	Antimicrobial activity	[73]
47.	*Sesbania grandiflora* (L.) Poiret	Leaf	22nm, Spherical	Cancer treatment	[74]
48.	*Tribulus terrestris* L.	Fruit	16–28 nm, spherical	Antimicrobial activity	[75]
49.	*Morinda citrifolia* L.	Root	30–55 nm, spherical	Cytotoxic	[76]
50.	*Phytolacca decandra* L.	Whole plant part	90.87 ± 0.005, spherical	Cytotoxicity through G2/M arrest in A375 cells	[77]

4. Mechanism of antibacterial inhibition by bioinspired Silver nanoparticles.

The actual mechanism of silver nanoparticles on the bacterial activity is not well understood; very meager work has been carried out in this aspect. The decrease in the AgNPs size can lead to an increase in the exact surface of a bactericidal specimen, suggests an increase in their ability to enter cell membrane, and thus improving antibacterial activity [78].

However, most recent literatures revealed that Ag^{+}, not the actual AgNPS, is the possible antimicrobial agent that causes cell damage with increase the Reactive oxygen species (ROS) cause cell damage and consequent to cell death [28,79], the free radical ion its derived from AgNPs surface its cause to cell death [80], it may also depletion of intracellular ATP concentration [81], it finally cause the respiratory enzymes in bacteria [82]. The possible mechanism of nanoparticles on the bacterial cell is displayed in Fig 1. The smaller the size of AgNPs more efficacies against the bactericidal activity as it provide a greater surface to the bacterial membrane.

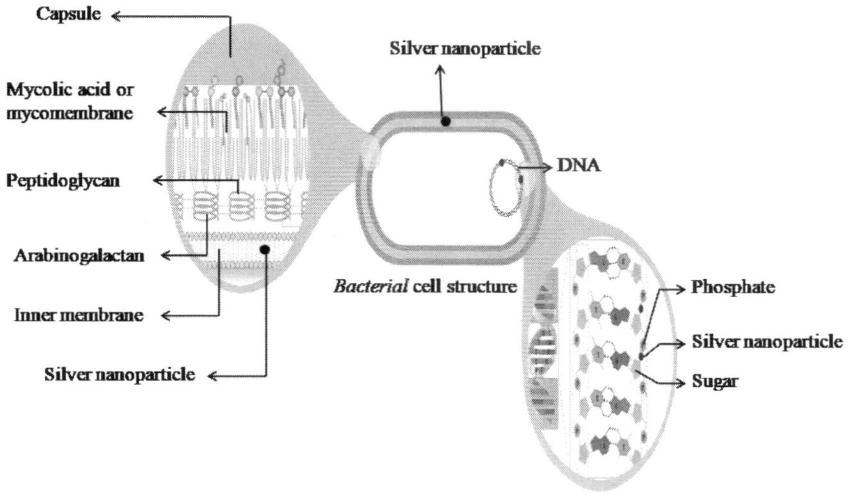

Graph 1: Number of cocoons produced per worm by Drawida modesta species

The activity is occurred between the positively charged Ag^+ ion with the negatively charged bacterial membrane, finally led to the disruption of the cell structure, leakage of cellular organelles occurred its resulting in cell death. In spite of these the AgNPS are strongly bind with the with phosphorus and sulfur of the extracellular and intracellular membrane proteins, these activity affects the cell replication, respiration and finally cell become death. The excellent bactericidal efficacy of AgNPs can be endorsed to the interaction with the plasma membrane and peptidoglycan cell wall of the bacterial strain [83-86]. It has also been suggested that the interaction of AgNPs with cell wall increases the membrane permeability by forming pores or pits and thereby causing the death of bacteria

Conclusion

Biogenic synthesis of silver nanoparticles using phytoextract shall be used as a potential alternative to chemical methods used for biomedical production of AgNPs due to its exceptionally high amount of reducing agents, which may not cause pruction of stable biogenic nanoparticles. This silver phytonanoparticles are one of the promising nanoproduct widely used in the field of bionanomedicine because of its unique properties. AgNPs having promised antimicrobial activity against bacteria, fungi, protozoa and certain viruses like antibacterial, antifungal, antiviral, antioxidant, anticancer, neuroprotective activity and other medical activity. Antibacterial properties of metallic silver and its derivatives are known from time of immemorial. In small concentrations, silver is safe for human cells but lethal for the majority of bacteria and viruses and hence widely used in disinfection of water and food in everyday life and in the infection control in medicine. Finally, the biogenic silver nanoparticles can improve the human health by less drug toxicity; these biomamatic nanoparticles may establish the pure green chemistry for environments.t.

Acknowledgement

The authors thanks to the Management, Department of Studies in Botany, Smt. Allum Sumangalamma Memorial College for Women, Ballari Karnatak, for providing the necessary facilities.

References

1. Frattini, A., Pellegri, N., Nicastro, D., Sanctis, OD ., 2005. Effect of amine groups in the synthesis of Ag nanoparticles using aminosilanes. Mater Chem Phys. 94,148–52.
2. Magudapathy P, Gangopadhyay P, Panigrahi B K, Nair KGM, Dhara, S. Electrical transport studies of Ag Nano clusters embedded in glass matrix. Physica B., 2001; 299(1-2): 142-146.
3. Joerger R, Klaus T, Granqvist CG. Biologicallyproduced silver-carbon composite materials for opticallyfunctional thin-film coatings. Adv Mater, 2000; 12(6): 407-409.
4. Panacek A, Kvitek L, Prucek R, Kolar M, Vecerova R, Pizurova N, et al. Silver colloid nanoparticles Synthesis, characterization, and their antibacterial activity. J Phys. Chem. B., 2006; 110(33): 16248-16253.
5. Mihailovic, V.; Sreckovic, N.; Nedic, Z.P.; Dimitrijevic, S.; Matic, M.; Obradovic, A.; Selakovic, D.; Rosic, G.; Katani Stankovic, J.S. Green Synthesis of Silver Nanoparticles Using *Salvia verticillata* and *Filipendula ulmaria* Extracts: Optimization of Synthesis, Biological Activities, and Catalytic Properties. Molecules 2023, 28, 808.
6. Dakshayani S.S., Marulasiddeshwara M.B., Sharath Kumar M.N., Ramesh Golla, Raghavendra Kumar P., Devaraja S., Rashmi Hosamani, Antimicrobial, anticoagulant and antiplatelet activities of green synthesized silver nanoparticles using *Selaginella* (Sanjeevini) plant extract, International Journal of Biological Macromolecules, 131, 2019, 787-797.
7. Wang, D., Xue, B., Wang, L. et al. Fungus-mediated green synthesis of nano-silver using Aspergillus sydowii and its antifungal/antiproliferative activities. Sci Rep 11, 10356 (2021)

8. Khan KU, Rehman M, Hameed M, Hayat K, Saeed F, Afzaal M, et al. (2025) Lantana camara plant extract catalyzed biosynthesis of graphene-metal nanocomposites with potential cytotoxic activities. PLoS ONE 20(3): e0314850.

9. Thyerre Santana da Costa, Gonzalo García Delgado, Carolyne Brustolin Braga, Ljubica Tasic, Insights into the fungal secretomes and their roles in the formation and stabilization of the biogenic silver nanoparticles. RSC Advances,15(9),2025,6938-6951.

10. Chika Iwuji, Hritaal Saha, William Ghann, Dominique Dotson, Md. Anwarul Kabir Bhuiya, Md. Shahnawaz Parvez, ZMG Sarwar Jahangir, Mohammed Muzibur Rahman, Faisal Islam Chowdhury, Jamal Uddin, Synthesis and characterization of silver nanoparticles and their promising antimicrobial effects, Chemical Physics Impact,9, 2024,100758

11. G. Velmathi, Velmurugan Sekar, Amutha Santhanam, Biogenic Synthesis of Silver Nanoparticles from Deinococcus radiodurans encapsulated with Curcumin for enhanced Antibiofilm activity, Microbial Pathogenesis, 2025, 107462, 0882-4010.

12. Stoyanova, M.; Milusheva, M.; Georgieva, M.; Ivanov, P.; Miloshev, G.; Krasteva, N.; Hristova-Panusheva, K.; Feizi-Dehnayebi, M.; Mohammadi Ziarani, G.; Stojnova, K.; et al. Synthesis, Cytotoxic and Genotoxic Evaluation of Drug-Loaded Silver Nanoparticles with Mebeverine and Its Analog. Pharmaceuticals 2025, 18, 397.

13. Zhang XF, Liu ZG, Shen W, Gurunathan S. Silver Nanoparticles: Synthesis, Characterization, Properties, Applications, and Therapeutic Approaches. Int J Mol Sci. 2016 Sep 13;17(9):1534.

14. [14] hika Iwuji, Hritaal Saha, William Ghann, Dominique Dotson, Md. Anwarul Kabir Bhuiya, Md. Shahnawaz Parvez, ZMG Sarwar Jahangir, Mohammed Muzibur Rahman, Faisal Islam Chowdhury, Jamal Uddin, Synthesis and characterization of silver nanoparticles and their promising antimicrobial effects, Chemical Physics Impact,9, 2024,100758.

15. Pathan, R., Vidyavathi, M., Kumar, R.V.S. et al. Design, Development, and Evaluation of Silver Nanoparticles Synthesized via Green Technology for Antioxidant, Antimicrobial, and Anticancer Activities. BioNanoSci. 15, 255 (2025).

16. Anita Dhaka, Suresh Chand Mali, Sheetal Sharma, Rohini Trivedi, A review on biological synthesis of silver nanoparticles and their potential applications, Results in Chemistry, 6, 2023,101108.

17. Silver S. Bacterial silver resistance: molecular biology and uses and misuses of silver compounds. FEMS Microbiol Rev. 2003; 27(2-3):341-53.

18. Chika Iwuji, Hritaal Saha, William Ghann, Dominique Dotson, Md. Anwarul Kabir Bhuiya, Md. Shahnawaz Parvez, ZMG Sarwar Jahangir, Mohammed Muzibur Rahman, Faisal Islam Chowdhury, Jamal Uddin, Synthesis and characterization of silver nanoparticles and their promising antimicrobial effects, Chemical Physics Impact,9,2024,100758.

19. Wozniak-Budych MJ, Staszak K, Staszak M. Copper and Copper-Based Nanoparticles in Medicine-Perspectives and Challenges. Molecules. 2023 Sep 18;28(18):6687.

20. Njud S. Alharbi, Nehad S. Alsubhi, Green synthesis and anticancer activity of silver nanoparticles prepared using fruit extract of *Azadirachta indica*, Journal of Radiation Research and Applied Sciences, 15(3),2022, 335-345.

21. Feoktistova M, Geserick P, Leverkus M. Crystal Violet Assay for Determining Viability of Cultured Cells. Cold Spring Harb Protoc. 2016; 1(4). pdb.prot087379.

22. Afrah Eltayeb Mohammed, Green synthesis, antimicrobial and cytotoxic effects of silver nanoparticles mediated by Eucalyptus camaldulensis leaf extract, Asian Pacific Journal of Tropical Biomedicine, 5(5),2015; 382-386.

23. Afshin, N., Mushtaq, N., Ahmed, M. et al. Therapeutic Potential of Green Synthesized Polyherbal Formulated Silver Nanoparticles in Alloxan-Induced Diabetes Mellitus—An In Vivo Strategy. Biol Trace Elem Res (2025).

24. Skoog, D.A., Holler, F.J. and Crouch, S.R. (2007) Principles of Instrumental Analysis. Sixth Edition, Brooks Cole, Belmont, 1039.

25. Chagas VT, França LM, Malik S, Paes AM. Syzygium cumini (L.) skeels: a prominent source of bioactive molecules against cardiometabolic diseases. Front Pharmacol. 2015 Nov 3;6:259.

26. Ponnusamy Kanchana, Venkatesan Hemapriya, Natarajan Arunadevi, Sankaran Shanmuga Sundari, Ill-Min Chung, Mayakrishnan Prabakaran, Phytofabrication of silver nanoparticles from Limonia acidissima leaf extract and their antimicrobial, antioxidant and its anticancer prophecy, Journal of the Indian Chemical Society,99(10), 2022.

27. Politano AD, Campbell KT, Rosenberger LH, Sawyer RG. Use of silver in the prevention and treatment of infections: silver review. Surg Infect (Larchmt). 2013;14(1):8-20.

28. Bheemanagouda N. Patil, T.C. Taranath, Limonia acidissima L. leaf mediated synthesis of silver and zinc oxide nanoparticles and their antibacterial activities, Microbial Pathogenesis, 115, 2018, 227-232.

29. Prasad T N V K V and Elumalai E K . Biofabrication of Ag nanoparticles using *Moringa oleifera* leaf extract and their antimicrobial activity. Asian Pacific Journal of Tropical Biomedicine.2011; 439-442 .

30. Daizy P and Unni C.Extracellular biosynthesis of gold and silver nanoparticles using krishna tulsi (*Ocimum sanctum*) leaf. Physica E: Low-dimensional Systems and Nanostructures.2011; 7: 1318–1322.
31. Pathipati U R and Pala R. Green synthesis of silver-protein (core–shell) nanoparticles using *Piper betle* L. leaf extract and its ecotoxicological studies on *Daphnia magna*. Colloids and Surfaces A: Physicochemical and Engineering Aspects . 2011; 1–3: 188–194.
32. Yilmaza M, Turkdemirb H, Akif K M, Bayramd E, Ciceke A, Metef A and Ulugg B. Biosynthesis of silver nanoparticles using leaves of *Stevia rebaudiana* . Materials Chemistry and Physics.2011; 130: 1195– 1202.
33. Mohsen Z, Azizah A H, Fatima A B, Mariana N S, Kamyar S, Fatemeh J and Farah F.Green synthesis and antibacterial effect of silver nanoparticles using *Vitex Negundo* L. Molecules.2011; 16: 6667-6676.
34. Rajendran K, Selvaraj M R, Arunachalam P, Venkatesan G K and Subhendu C. Agricultural waste *Annona squamosa* peel extract: Biosynthesis of silver nanoparticles . Spectrochimica Acta Part A . 2012; 90: 173– 176 .
35. Antariksh S, Tripathi R M, Fahmina Z and Priti S. Green synthesis of silver nanoparticles using aqueous solution of *Ficus benghalensis* leaf extract and characterization of their antibacterial activity. Materials Letters.2012; 67: 91–94.
36. Gole A, Dash C, Ramakrishnaan V, Sainkar S R, Mandal A B, Rao M and Sastry M. Pepsin–Gold Colloid Conjugates: Preparation, Characterization and Enzymatic Activity. Langmuir.2001; 17: 1674–1679.
37. Happi E T, Robert C, Ronkart S N, Wathelet B and Paquot M. Dietary fibre components and pectin chemical features of peels during ripening in banana and plantain varieties. Bioresour. Technol. 2007; 99: 4346–4354.
38. Quirantes-Piné R, Funes L , Micol V, Segura-Carretero A and Fernández-Gutiérrez, A. High-performance Liquid Chromatography with Diode Array Detection Coupled to Electrospray Time-of-flight and Ion-trap Tandem Mass Spectrometry to Identify Phenolic Compounds From a Lemon Verbena Extract. J Chromatogr A . 2009; 28: 5391-5397.
39. Gruen L C. Interaction of amino acids with silver ions. Biochim. Biophys. Acta.1975; 386: 270–274.
40. Chen P S, Li J H, Liu T Y, Lin T C. Folk medicine *Terminalia catappa* and its major tannin component, Punicalagin, are effective against bleomycin-induced genotoxicity in Chinese Hamster ovary cells. Cancer Letters. 2000; 152: 115–122.
41. Kawsas S M A, Huq E, Nahar N and Ozeki Y. Identification and quantification of phenolic acids in *Macrotyloma uniflorum* by reversed phase-HPLC. American Journal of Plant Physiology. 2008; 3: 165–172.

42. Comell D W and McLachlan R. Natural pungent compounds: IV. Examination of the gingerols, shogaols, paradols and related compounds by thin-layer and gas chromatography. J Chromatogr .1972; 67 :29–35.

43. [44] Tajammul H S, Iqbal M and Mazhar M. Size control synthesis of starch capped-gold nanoparticles. J Nanopart Res.2009; 11: 1383–1391.

44. Madhavaraj Lavanya1, Sethumadhavan Vishnu Veenavardhini1, Geun Ho Gim2, Mathur Nadarajan Kathiravan and Si Wouk Kim. Synthesis, Characterization and Evaluation of Antimicrobial Efficacy of Silver Nanoparticles using *Paederia foetida* L. leaf extract. *International Research Journal of Biological Sciences* Vol. 2(3), 28-34, March (2013).

45. Garima Sharma 1,2,†,*, Ashish Ranjan Sharma 2,†, Riju Bhavesh 1, Jongbong Park 2, Bilguun Ganbold 2, Ju-Suk Nam 2 and Sang-Soo Lee. Biomolecule-Mediated Synthesis of Selenium Nanoparticles using Dried *Vitis vinifera* (Raisin) Extract. *Molecules* 2014, *19*, 2761-2770.

46. R. Mohammadinejad1*, Sh. Pourseyedi2, A. Baghizadeh3, Sh. Ranjbar1, G. A. Mansoori4*. Synthesis of Silver Nanoparticles Using Silybum Marianum Seed Extract.Int. J. Nanosci. Nanotechnol., Vol. 9, No. 4, Dec. 2013, pp. 221-226.

47. K. Vinothkumar* C. Udayasoorian and R.M. Jayabalakrishnan. Biobase Synthesis of Silver Metal Nanoparticles Using *Murraya koenigii* Leaf Extract. *Int. J. Pure App. Biosci.* 2 (2): 244-248 (2014).

48. Sohail Yasin a, Lin Liu a, Juming Yao. Biosynthesis of Silver Nanoparticles by Bamboo Leaves Extract and Their Antimicrobial Activity. Journal of Fiber Bioengineering and Informatics 6:1 (2013) 77-84.

49. R Muthukumar1, R Chidambaram2, and V Ramesh3. Biosynthesis of Silver Nanoparticles from *E.Tirucalli* and to Check Its Antimicrobial Activity. 2014 RJPBCS 5(2). 589-596.

50. B. Sundararajan1 and B. D. Ranjitha Kumari2. Biosynthesis of silver nanoparticles in *Lagerstroemia speciosa* (l.) pers and their antimicrobial activities. Int J Pharm Pharm Sci, Vol 6, Issue 3, 30-34.

51. Himakshi Bhati-Kushwaha and CP Malik. Biosynthesis of silver nanoparticles using fresh extracts of *Tridax procumbens* Linn. Indian journal of experimental biology. Vol. 52, 2014, pp. 359-368.

52. Akl M. Awwad1*, Nidà M. Salem2, Amany O. Abdeen. Biosynthesis of silver nanoparticles using Loquat leaf extract and its antibacterial activity. Adv. Mat. Lett. 2013, 4(5), 338-342.

53. P. Shivakumar Singh and G. M. Vidyasagar. Biosynthesis, Characterization, and Antidermatophytic Activity of Silver Nanoparticles Using Raamphal Plant (*Annona reticulata*) Aqueous Leaves Extract.Indian Journal of

Materials Science. Volume 2014 (2014), Article ID 412452, 5 pages.

54. I.B. Gokak and T.C. Taranath. *Cansjera rheedii* J. F. Gmel. a medicinal plant-mediated synthesis of Silver nanoparticles and their antibacterial activity. International Journal of Scientific Engineering and Technology. 3 (3): 293 – 296.

55. Manikandan Ramar, Beulaja Manikandan, Prabhu Narayanan Marimuthu, Thiagarajan Raman, Anjugam Mahalingam, Palanisamy Subramanian, Saravanan Karthick, Arumugam Munusamy, Synthesis of silver nanoparticles using Solanum trilobatum fruits extract and its antibacterial, cytotoxic activity against human breast cancer cell line MCF 7, Spectrochimica Acta Part A: Molecular and Biomolecular Spectroscopy,140, 2015, 223-228,

56. Akl M Awwad, Nida M Salem and Amany O Abdeen. Green synthesis of silver nanoparticles using carob leaf extract and its antibacterial activity. International Journal of Industrial Chemistry 2013, 4:29 1-6.

57. Harekrishna Bar, Dipak Kr. Bhui, Gobinda P. Sahoo, Priyanka Sarkar, Sankar P. De, Ajay Misra, Green synthesis of silver nanoparticles using latex of *Jatropha curcas*, Colloids and Surfaces A: Physicochemical and Engineering Aspects, 339, 1–3, 2009, 134-139.

58. Ageng Azhar Amar, Effendy De Lux Putra, Denny Satria, Panal Sitorus, Hooi Ling Lee, Syukur Berkat Waruwu, Mahatir Muhammad, Biosynthesis And Biological Activity Of Silver Nanoparticles From Zanthoxylum Acanthopodium Fruits, South African Journal of Chemical Engineering, 2025,

59. Saud B Green synthesis of silver nanoparticles from Euphorbia milii plant extract for enhanced antibacterial and enzyme inhibition effects International Journal of Health Sciences Vol. 18, Issue 2; 2024.

60. T.C. Prathna, N. Chandrasekaran, Ashok M. Raichur, Amitava Mukherjee, Biomimetic synthesis of silver nanoparticles by Citrus limon (lemon) aqueous extract and theoretical prediction of particle size, Colloids and Surfaces B: Biointerfaces, 82, Issue 1, 2011, 152-159.

61. Mukunthan K S, Elumalai E K, Trupti N P and Ramachandra M V. *Catharanthus roseus*: A natural source for the synthesis of silver nanoparticles. Asian Pacific Journal of Tropical Biomedicine. 2011; 1: 270-274.

62. Hambardzumyan S, Sahakyan N, Petrosyan M, Nasim MJ, Jacob C, Trchounian A. Origanum vulgare L. extract-mediated synthesis of silver nanoparticles, their characterization and antibacterial activities. AMB Express. 2020 5;10(1):162.

63. Sathishkumar, M. Krishnamurthy, S. Yun, Y. S. Immobilization of silver nanoparticles synthesized using the Curcuma longa tuber powder extract on cotton cloth for bactericidal activity. Biores. Technol. 2010, 101 (20), 7958 7965.

64. Ravichandran Veerasamy, Tiah Zi Xin, Subashini Gunasagaran, Terence Foo Wei Xiang, Eddy Fang Chou Yang, Nelson Jeyakumar, Sokkalingam Arumugam Dhanaraj, Biosynthesis of silver nanoparticles using mangosteen leaf extract and evaluation of their antimicrobial activities, Journal of Saudi Chemical Society, 15(2),2011,113-120

65. [66] Rupali S. Patil, Mangesh R. Kokate, Sanjay S. Kolekar, Bioinspired synthesis of highly stabilized silver nanoparticles using Ocimum tenuiflorum leaf extract and their antibacterial activity, Spectrochimica Acta Part A: Molecular and Biomolecular Spectroscopy, 91,2012,234-238.

66. Kumkoon T, Srisaisap M, Boonserm P. Biosynthesized Silver Nanoparticles Using Morus alba (White Mulberry) Leaf Extract as Potential Antibacterial and Anticancer Agents. Molecules. 2023 Jan 26;28(3):1213.

67. Elakuu Tesfaye, Yodahe Gonfa, Getachew Tadesse, Tatek Temesgen, Selvakumar Periyasamy, Green synthesis of silver nanoparticles using Vernonia amygdalina plant extract and its antimicrobial activities, Heliyon, 9(6),2023, e17356.

68. Daizy P and Unni C. Extracellular biosynthesis of gold and silver nanoparticles using krishna tulsi (*Ocimum sanctum*) leaf. Physica E: Low-dimensional Systems and Nanostructures. 2011; 7: 1318–1322.

69. Kamaraj C, Ragavendran C, Manimaran K, Sarvesh S, Islam ARMT, Malafaia G. Green synthesis of silver nanoparticles from Cassia Auriculata: Targeting antibacterial, antioxidant activity, and evaluation of their possible effects on saltwater microcrustacean, Artemia Nauplii (non-target organism). Sci Total Environ. 2023 25;861:160575.

70. Govender R, Phulukdaree A, Gengan RM, Anand K, Chuturgoon AA. Silver nanoparticles of Albizia adianthifolia: the induction of apoptosis in human lung carcinoma cell line. J Nanobiotechnology. 2013 Feb 18;11:5. doi: 10.1186/1477-3155

71. Murugaraj Jeyaraj, Manoharan Rajesh, Renganathan Arun, Davoodbasha MubarakAli, Gnanasekar Sathishkumar, Ganeshan Sivanandhan, Gnanajothi Kapil Dev, Markandan Manickavasagam, Kumpati Premkumar, Nooruddin Thajuddin, Andy Ganapathi, An investigation on the cytotoxicity and caspase-mediated apoptotic effect of biologically synthesized silver nanoparticles using Podophyllum hexandrum on human cervical carcinoma cells, Colloids and Surfaces B: Biointerfaces, 102, 2013,708-717.

72. Sivaramakrishnan, M., Jagadeesan Sharavanan, V., Karaiyagowder Govindarajan, D. et al. Green synthesized silver nanoparticles using aqueous leaf extracts of Leucas aspera exhibits antimicrobial and catalytic dye degradation properties. SN Appl. Sci. 1, 208 (2019).

73. Rengarajan, S., Thangavel, N., Sivalingam, A.M. et al. Green synthesis and characterization of silver nanoparticles with different solvent extracts of Sesbania grandiflora (L.) Poiret and assessment of their antibacterial and antioxidant potentials. Biomass Conv. Bioref. (2023).

74. V. Gopinath, D. MubarakAli, S. Priyadarshini, N. Meera Priyadharsshini, N. Thajuddin, P. Velusamy, Biosynthesis of silver nanoparticles from Tribulus terrestris and its antimicrobial activity: A novel biological approach, Colloids and Surfaces B: Biointerfaces, 96, 2012,69-74.

75. Suman TY, Radhika Rajasree SR, Kanchana A, Elizabeth SB. Biosynthesis, characterization and cytotoxic effect of plant mediated silver nanoparticles using Morinda citrifolia root extract. Colloids Surf B Biointerfaces. 2013 Jun 1;106:74-8.

76. Sreemanti Das, Jayeeta Das, Asmita Samadder, Soumya Sundar Bhattacharyya, Durba Das, Anisur Rahman Khuda-Bukhsh, Biosynthesized silver nanoparticles by ethanolic extracts of Phytolacca decandra, Gelsemium sempervirens, Hydrastis canadensis and Thuja occidentalis induce differential cytotoxicity through G2/M arrest in A375 cells, Colloids and Surfaces B: Biointerfaces, 101, 2013, 325-336

77. Baker C, Pradhan A, Pakstis L, Pochan DJ, Shah SI (2005) Synthesis andantibacterial properties of silver nanoparticles. J NanosciNanotechnol 5:244–249

78. Xu, H., Qu, F., Xu, H. et al. Role of reactive oxygen species in the antibacterial mechanism of silver nanoparticles on Escherichia coli O157:H7. Biometals 25, 45–53 (2012).

79. McQuillan JS, Infante HG, Stokes E, Shaw AM. Silver nanoparticle enhanced silver ion stress response in Escherichia coli K12. Nanotoxicology. 2012 Dec;6:857-66.

80. Mahendra Rai, Alka Yadav, Aniket Gade, Silver nanoparticles as a new generation of antimicrobials, Biotechnology Advances, 27(1), 2009, 76-83.

81. Patil, M.P., Kim, GD. Eco-friendly approach for nanoparticles synthesis and mechanism behind antibacterial activity of silver and anticancer activity of gold nanoparticles. Appl Microbiol Biotechnol 101, 79–92 (2017).

82. Ivan Sondi, Branka Salopek-Sondi, Silver nanoparticles as antimicrobial agent: a case study on E. coli as a model for Gram-negative bacteria, Journal of Colloid and Interface Science, 275(1), 2004, 177-182.

83. Ivan Sondi, Branka Salopek-Sondi, Silver nanoparticles as antimicrobial agent: a case study on E. coli as a model for Gram-negative bacteria, Journal of Colloid and Interface Science, 275(1), 2004, 177-182.

84. Dabin Yu, Vivian Wing-Wah Yam Controlled Synthesis of Monodisperse Silver Nanocubes in Water J. Am. Chem. Soc. 2004, 126, 41, 13200–13201.

85. Chhangte Vanlalveni, Samuel Lallianrawn, Ayushi Biswas, Manickam Selvaraj, Bishwajit Changmai and Samuel Lalthazuala Rokhum, Green synthesis of silver nanoparticles using plant extracts and their antimicrobial activities: a review of recent literature. RSC Adv., 2021,11, 2804-2837

CHAPTER 12

DIVERSITY OF *DROSOPHILA* SPECIES IN AND AROUND GAVIMUTT CAMPUS, KOPPAL: A FIELD STUDY

Prashanth Konkal[1], Shwetha G[1], Bheemanagouda N.P[2], and Channabasava S[1]

1Shri Gavisiddeshwar Arts, Science and Commerce College, Koppal – 584231
2Department of Botany, Smt. ASM College for Women Ballari, Karnataka, India,
Correspondence: Prashanth Konkal (Konkalprashanth50@gmail.com)

ABSTRACT:

Drosophila species, commonly known as fruit flies, are an excellent model for studying genetics, behavior, and ecology. This study investigates the diversity and distribution of Drosophila species in and around SG College Campus, Koppal. Collections were made using traps with apple vinegar attractants and net sweeps. A total of five Drosophila species were identified: *Drosophila bipectinata* (40%), *Drosophila nasuta* (25%), *Drosophila annanassae* (20%), *Drosophila malara* (10%), and *Drosophila neo-nasuta* (5%). The Shannon-Wiener Diversity Index (H' = 1.415) and Simpson's Diversity Index (D = 0.725) indicate a moderate level of species diversity. Environmental factors such as temperature, humidity, and food availability influence species composition. Conservation efforts are necessary to sustain species diversity in urbanized areas.

Introduction

Drosophila, commonly known as fruit flies, are an excellent model organism for studying genetics, behavior, and ecology (Bashisth & Singh, 2015). The genus Drosophila belongs to the family Drosophilidae (Class: Insecta, Order: Diptera) and is known for its diverse range of species, with over 1,500 species found worldwide (Markow & O'Grady, 2006). The name Drosophila is derived from the Greek words "drosos" meaning dew and "phila" meaning loving, which refers to their affinity for moist environments (Guruprasad & Padmaja, 2016). Drosophila species are found in a wide range of habitats, including tropical and temperate regions, and are known to be highly sensitive to environmental changes (Srinath & Shivanna, 2014).

The study of Drosophila diversity is important for understanding the ecological and evolutionary processes that shape their populations (Parshad et al., 1980). In India, the diversity of Drosophila species is poorly studied, with only a few studies conducted in the past (Dwivedi & Gupta, 1980). However, recent studies have shown that the Western Ghats region of Karnataka is home to a diverse range of Drosophila species (Krishnan & Koushik Ponnanna, 2012). The present study aims to investigate the diversity of Drosophila species in and around the SG College Campus, Koppal, and to contribute to the existing knowledge on the distribution and ecology of Drosophila species in this region. The SG College Campus, Koppal, provides a unique opportunity to study the diversity of Drosophila species in a semi-urban environment (Nituraj & Krishnan, 2015). The campus is located in a region with a dry climate, which is expected to support a diverse range of Drosophila species (Guruprasad & Padmaja, 2016). The study of Drosophila diversity in this region is important for understanding the ecological and evolutionary processes that shape their populations, and for developing strategies for conservation and management of these species.

Materials and Methods

The drosophila is highly versatile organisms for carrying out laboratory studies as they are easily procurable and cannot convenient to handle. Initially many collection sites were spotted in and around Koppal but due to uneven sizes and mixed species composition, one fish site was chosen to collect the drosophila for the pilot study.

Collection site

The drosophila was collected by using trap with an attractant like apple vinegar, sticky nets, around the site located in Koppal, Gavimath campus, Karnataka shown in figure 3.

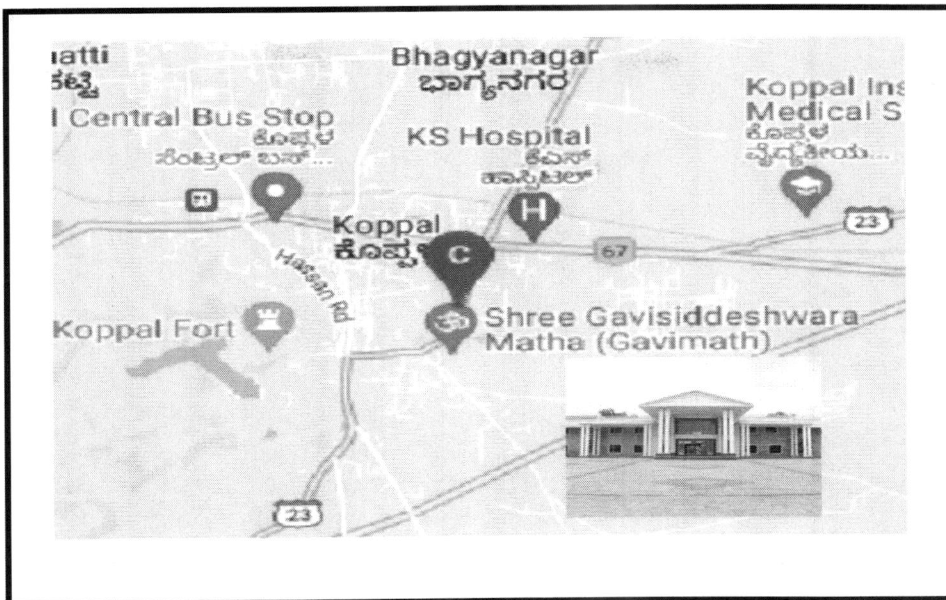

Figure 1: Shri Gavisiddeshwara Arts, Science and commerce college campus, Koppal

Collection Methods

The following steps were used to collect the species.

1. Installing trap bait

Fruit slices are used as bait in box these traps kept under a tree. The collection area must be damp and moist with minimal human interference. After 1-2 days traps must be recovered to collect flies, same method could be applied to collect all kind of drosophila species.

2. Net swipe

There are many genera that are not attached to regular traps and need to be captured from their natural food source (wild rotting fruits).capturing such drosophila species nets are most effective. The flies hovering over the rotten fruits or piles of organic waste could be captured using this method.

3. Media preparation

Required quantities of water are taken in a pressure cooker. Corn flour, glucose, sugar, yeast powder and agar are added to a boiled water. close the pressure cooker with a lid without cooker weight and keep on a high heat for a

5-min.open the cooker and mix thoroughly, cool the media. The media is used as a trap to catch fruit flies.

Result

Identification of Drosophila species

Samples of drosophila species were collected from selected sites.

The Sex determination of drosophila were identified on the basis of morphological characters like,

- The basic character to find the sex of drosophila is with the color of eyes and types of wing present in it.
- Abdomen Structure: The males have pointed abdomen and the females have blunt abdomen.

The males have sex combs in the legs and number of teeth and pattern and females has ovipositor in abdomen region. The adult drosophila body is compressed of three major parts are Head, Thorax and Abdomen

The drosophila has one pair of wings and halters each, wings help in flight whereas halters help as balancing organ-

1. Head: Head of drosophila has distinguishing parts like ocellar triangle, post oceller setae inner and outer vertical setae. Fronto-orbital plates cibarium and arista
2. Thorax: The thorax of the drosophila consists of three segments Prothorax, Mesothorax, Metathorax. These are moving from anterior to posterior, each segment possesses a pair of legs. The wings attach to the mesothorax.
3. Abdomen: Contains seven pairs of holes (spiracles), one pair in each segment. The eighth and last segment of the female contains the anus and no spiracles. In the female, the vulva lies between the apices of the gonopods, these irregular tubes extend into the body from the surface.

1. *Drosophila bipectinata*

Classification: Kingdom: Animalia

Phylum: Arthopoda

Class: Insecta

Order: Diptera

Family: Drosophilidae

Genus: Drosophila

Species: D. *bipectinata*

Characters:

- The abdomen coloration brown with black stripes.
- In D. bipectinata male, first tarsmere has a teeth in proximal comb, and distal comb.
- 11 sex combs teeth arrangement pattern have been detected in D.bipectinata and sex comb teeth number ranges from 10-15.thus sex comb teeth number and arrangement varied most in D. bipectinata compared to other species

Distribution: D. bipectinata was first discovered by Duda from Darjeeling, India. D. bipectinata is distributed in the oriental region and known from Reunion Island. These species occur in Karnataka in India. D. bipectinata is most widely distributed. Females are indistinguishable, but males are distinguishable by their sex-comb teeth number and pattern and by abdominal coloration. Chromosomal inversions have been detected in these species. In natural populations of D. bipectinata the frequency of inversions and the level of inversion heterozygosity were found to be very low but in laboratory stocks inversions persisted for more than 20 generations due to heterotic buffering. Taxonomically, drosophila bipectinata species complex are the members of sub genus sophophora of melanogaster species group. Female are indistinguishable and no differences have been found in Genitalia of male or of females. Males are distinguished however by their sex comb teeth number and pattern and also by abdomen coloration.

2. ***Drosophila nasuta***

Classification: Kingdom: Animalia

Phylum: Arthopoda

Class: Insecta

Order: Diptera

Family: Drosophilidae

Genus: Drosophila

Species: D. *nasuta*

Characters:

- Head, anti dorsal view, showing the conspicuous iridescent white silvery forms.
- Body showing the large brownish stripe on half dorsal area of pleura.
- The abdomen length is about 1mm.
- The number of sex- combs is present in the Drosophila nasuta species
- In the region of the species Drosophila nasuta is present.
- The female has Ovipositor and male have the number of sex-combs and based on abdomen patterns the differentiation of sex is obtained.

3. *Drosophila annanassae*

Classification: Kingdom: Animalia

Phylum: Arthopoda

Class: Insecta

Order: Diptera

Family: Drosophilidae

Genus: Drosophila

Species: D. *annanassae*

Characters:

- In D. annanassae the abdomen coloration is light brown with strips.
- The 8 rows of sex comb present.
- Total sex comb teeth number detected in D. annanassae is 32-33.
- Only 2 patterns of sex comb teeth arrangement were found.
- Sex comb teeth number in D. annanasae varies from 22-35 to 31-35.

4. *Drosophila malara*

Classification: Kingdom: Animalia

Phylum: Arthopoda

Class: Insecta

Order: Diptera

Family: Drosophilidae

Genus: Drosophila

Species: D. *malhara*

Characters:

- D. malhara can live in temperature and available of water. In temperate regions where human activities have introduced D. malhara, these flies seek shelter in a colder winter months.
- D. malhara is a simplistic. They are easily drawn towards the smell of any food source

5. ***Drosophila neo-nasuta***

Classification: Kingdom: Animalia

Phylum: Arthopoda

Class: Insecta

Order: Diptera

Family: Drosophilidae

Genus: Drosophila

Species: D. *neo-nasuta*

Characters:

- The males of D. neo-nasuta have 2 obliquely placed sex combs on metatarsal segment, upper comb with 5-6 teeth's.
- Lower comb with 6-9 teeth's and 1-2 teeth on a tarsal segment of first segment.

Statistical Analysis

Shannon-Wiener Diversity Index (H') = 1.415

This indicates a moderate level of species diversity in the study area. Higher values (closer to 2 or 3) indicate greater species diversity, while lower values (~1) suggest dominance by a few species. In this case, D. bipectinata is dominant, leading to a slightly lower diversity index.

Simpson's Diversity Index (D) = 0.725

This measures species dominance (closer to 1 means higher diversity, closer to 0 means a single species dominates). A value of 0.725 suggests that while *D. bipectinata* is dominant, there is still a reasonable level of species diversity.

Results and Discussion

Table 1: Showing the percentage of Drosophila species distribution in the study area.

Sl. No	Name of the species	Percentage
1	Drosophila bipectinata	40%
2	Drosophila nasuta	25%
3	Drosophila annanassea	20%
4	Drosophila malara	10%
5	Drosophila neo-nasuta	05%

Table: 1. Percentage of drosophila species found in campus

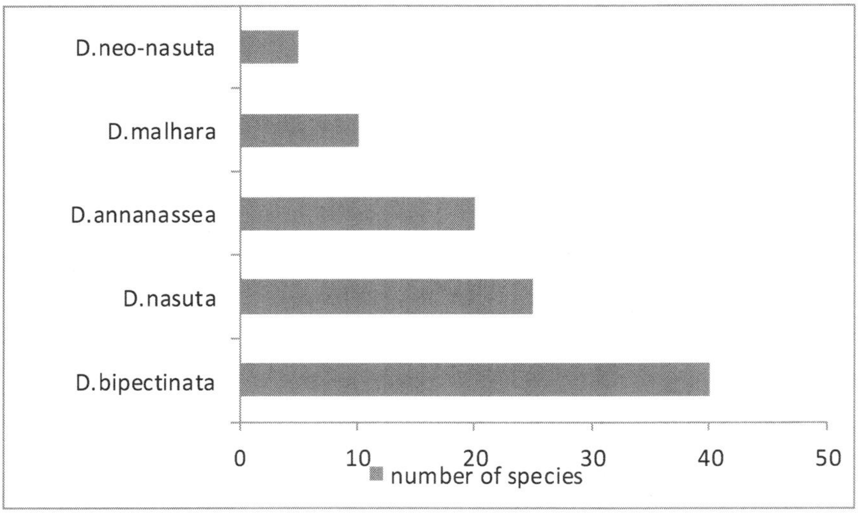

Fig.2 Bar chart showing the drosophila species around campus

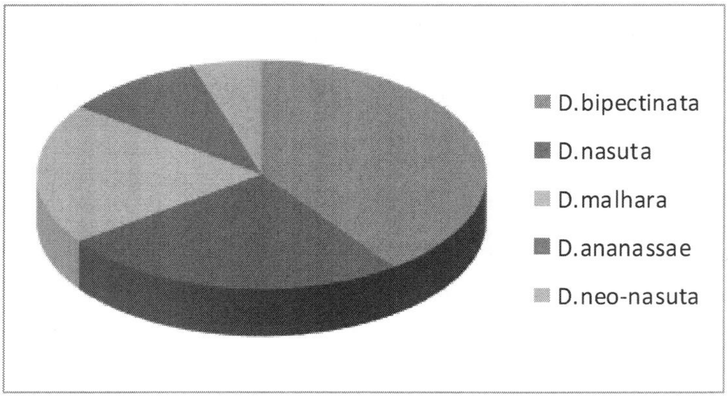

Fig.3: Pie chart showing the drosophila species around campus

The present study investigated the diversity and distribution of *Drosophila* species in and around SG College Campus, Koppal, revealing the presence of five species with varying levels of abundance. The findings indicate that *Drosophila bipectinata* was the most dominant species, representing 40% of the total population, followed by *Drosophila nasuta* (25%), *Drosophila annanassae* (20%), *Drosophila malara* (10%), and *Drosophila neo-nasuta* (5%). This variation in species distribution suggests that environmental factors such as temperature, humidity, food availability, and habitat structure play a crucial role in shaping local Drosophila populations. A pie chart representing the species composition provides a clear visual distribution of these species. The chart highlights the dominance of *D. bipectinata,* which occupies the largest proportion, followed by moderately abundant species such as *D. nasuta* and *D. annanassae.* The relatively smaller portions representing *D. malara* and *D. neo-nasuta* indicate their limited presence in the study area, potentially due to competition, habitat preferences, or other ecological constraints. The bar chart further reinforces these trends by allowing a comparative visualization of the population sizes. The tallest bar, representing *D. bipectinata,* illustrates its overwhelming dominance, while progressively shorter bars for other species emphasize their decreasing population sizes. Such graphical representations facilitate a better understanding of biodiversity patterns in the study area.

To assess species diversity statistically, Shannon-Wiener and Simpson's Diversity Indices were calculated. The Shannon-Wiener Diversity Index (H' = 1.415) suggests moderate species diversity, meaning that while *D. bipectinata* is the most common species, other species coexist within the habitat. Similarly, Simpson's Diversity Index (D = 0.725) indicates that no single species dominates entirely, suggesting a reasonably balanced ecosystem. These findings align with previous research demonstrating that in tropical and subtropical climates, certain *Drosophila* species tend to dominate due to their ecological adaptability and reproductive efficiency. The dominance of *D. bipectinata* in the study area may be attributed to its high reproductive potential, ability to efficiently utilize available food resources, and adaptability to local climatic conditions. Studies have shown that *D. bipectinata* thrives in warm, humid environments where fermenting fruits and organic matter are readily available, making it a highly competitive species in such habitats. On the other hand, species such as *D. neo-nasuta,* which accounted for only 5% of the total population, may require specific microhabitat conditions that are less prevalent in the study area.

The presence of *D. nasuta* (25%) and *D. annanassae* (20%) in substantial proportions suggests that these species are competitively strong but are not the dominant species in this particular habitat. Research indicates that *D. nasuta* is well-adapted to disturbed and semi-urban environments, allowing it to coexist alongside dominant species. Conversely, the lower representation of *D. malara* and *D. neo-nasuta* suggests that they might be less competitive or more specialized in habitat requirements, limiting their population size. Environmental factors, including habitat structure, availability of food resources, and interspecies competition, significantly influence the distribution of *Drosophila* species. The study area, being a semi-urban environment, provides moderate vegetation cover and human-induced ecological modifications, which could favor certain species over others. Previous studies suggest that urban and semi-urban settings often lead to a decline in species diversity, allowing generalist species to dominate while specialists decline. In this context, the high proportion of *D. bipectinata* and *D. nasuta* may be indicative of their ability to adapt to anthropogenic changes, whereas the relatively lower abundance of *D. neo-nasuta* may suggest a preference for undisturbed natural habitats. To further evaluate the significance of species distribution, a Chi-square goodness-of-fit test could be applied to determine whether the observed variation is statistically significant. If the test yields a p-value < 0.05, it would confirm that the species distribution is not random and is instead influenced by specific environmental or biological factors. Given the dominance of *D. bipectinata* and the differing proportions of other species, it is likely that such an analysis would show significant variation in species distribution patterns.

These results have important ecological implications, particularly in understanding species interactions, competitive exclusion, and habitat preferences. The findings suggest that certain species outcompete others in specific environmental conditions, leading to an uneven species distribution. This pattern is consistent with previous research, which indicates that some *Drosophila* species thrive in highly competitive environments, while others are restricted to specialized ecological niches. Additionally, the presence of moderate diversity suggests that although one species is dominant, the habitat still supports a range of species, preventing complete ecological monopolization. Future research should focus on seasonal variations in *Drosophila* species diversity, as different environmental conditions throughout the year could influence population dynamics. Additionally, molecular and genetic studies could provide insights into the evolutionary relationships between these species, helping to understand adaptation mechanisms and genetic variations. Further investigations into climate change and habitat modifications would also be valuable in assessing long-term shifts in species diversity and predicting future trends in *Drosophila* populations.

In conclusion, this study provides a comprehensive analysis of *Drosophila* species diversity in SG College Campus, Koppal. The findings reveal that while *D. bipectinata* is the most dominant species, the habitat still supports moderate biodiversity, allowing multiple species to coexist. The statistical and graphical analysis further confirms the unequal species distribution, highlighting potential ecological interactions and environmental influences. These insights contribute to a broader understanding of species adaptation, competition, and ecosystem stability in semi-urban environments. Future studies should incorporate longitudinal observations, genetic analysis, and climate modeling to deepen our understanding of Drosophila population dynamics and their ecological significance.

Acknowledgement

We acknowledge the SGVVTs management for supporting this research.

References

1. Bashisth, N., & Singh, B. N. (2015). Species and genetic diversity in genus Drosophila inhabiting the Indian sub-continent. Journal of Genetics, 94(3), 351-356.
2. Markow, T. A., & O'Grady, P. M. (2006). Drosophila biology: a guide to species, genes, and methods. Academic Press.
3. Guruprasad, B. R., & Padmaja, C. (2016). Drosophila – a model organism for assessment of biodiversity. Journal of Entomology and Zoology Studies, 4(3), 88-90.
4. Srinath, B. S., & Shivanna, N. (2014). Seasonal variation in natural population of Drosophila in Dharwad, India. Journal of Entomology and Zoology Studies, 2(4), 35-41.
5. Parshad, R., Paika, I. J., & Hegde, S. N. (1980). Study on Drosophila nasuta sub-group on distribution and morphology. Journal of the Bombay Natural History Society, 77(2), 241-248.
6. Dwivedi, Y. N., & Gupta, J. P. (1980). Study on subgenus of Drosophila. Journal of the Bombay Natural History Society, 77(1), 34-41.
7. M.S. Krishnan, & Koushik Ponnanna, C. R. (2011-2012). Biodiversity of Drosophila of Western Ghats of Karnataka. Journal of Entomology and Zoology Studies, 95, 1-8.
8. Nituraj, & M.S. Krishnan. (2015). Contribution to the behavioral genetics of Drosophila bipectinata complex. Journal of Genetics, 94(2), 257-264.

INDEX